Open(

The innovator's guide to the Mobile data industry

By
Ajit Jaokar and Tony Fish

Copyright 2004 futuretext Limited
Issue Date 15 November 2004
Published by
futuretext
36 St George St
Mayfair
London
W1S 2FW, UK
Email:info@futuretext.com
www.futuretext.com

ISBN:0-9544327-2-X

Table of Contents

Foreword

Where have all the flowers gone? - the mobile innovation, ideas, investment, partners, entrepreneurs and growth prospects? This book explains the barriers to innovation in the mobile data industry and ways to address these opportunities. It offers independent insight to developers and investors without technical bias.

For those of us who believe in a multi-access, multi-network world of mobile multimedia, applications anywhere anytime, the book provides a focus on the delivery of mobile applications that is refreshing. It offers illustrations and experience in how to develop and deliver mobile applications for different markets, without excessive technical jargon. As developers repurpose for the 4th screen of mobile, from the origination for the 2nd or 3rd screen of TV or PC`s, then this guide will provide a useful backdrop. The flow of Liquid Media for customers in this multi access world, across these traditionally separate industries, will be enhanced by this material. The Content / services of the Mobile shopping malls or mobile portals, will also be enriched as we seek access to this content from more pocket mobiles / pda`s than ever before.

As a reference guide it highlights some of the challenges to be faced by developers in the delivery, partnership and exploitation of mobile content and applications. It is not a design handbook, but talks openly of business models and relevant design guidance for such mobile development. It also offers a useful set of web references for the reader who needs more.

It is also a "roadmap" of standards and applications coupled with some of the key areas for developer focus. But OpenGardens is also different, as the book adopts a more philosophical approach to opening up mobile "gardens" for wider use and enjoyment. It also offers an open approach to web based follow up. This book provides the scent to guide the reader to these gardens, in order that these "flowers" and applications flourish and grow.

Mike Short
Vice President, Research and Development - O2 GROUP
Chairman - Mobile Data Association
Past Chairman - GSM Association

Preface

OpenGardens draws on our interaction with visionary developers and innovators in the Mobile data industry including the 5,000 people in our networks and on also our personal experience. It provides a down to earth approach for the 'practical innovator', but has a deep running cause. The book presents ideas that we believe deeply in – within an industry we understand at the grassroots.

Throughout this book, we use the word 'developer' interchangeably with an 'innovator'. Our aim is to help the innovator bring their Mobile services to market and help them gain a fair reward. Based on our findings, we believe that the third party developer ecosystem in the Mobile data industry is broken. This book is not a deep strategic understanding of why it is broken or how to fix it. Rather, it is about how developers and Mobile operators can work together to create new wealth even within a broken ecosystem. Recognising that there are problems in the existing ecosystem, we propose a model that will work even with the broken processes. With viewpoints from both developers and Mobile operators, OpenGardens is all about growing a new ecosystem at some point in the future – one in which each side benefits whilst at the same time – making the best of the existing ecosystem today.

The authors, Ajit and Tony, are uniquely qualified to write OpenGardens. Between them, they offer a dual perspective covering development knowledge, telecoms and financing. Through their online and offline networking forums, Ajit and Tony have worked with a wide range of entrepreneurs. Ajit brings to the book the telecoms/developer perspective whereas Tony brings the Investor/finance perspective. Over the last year, through our networks, we have interacted with entrepreneurs from all over the world – spanning more than 91 countries ranging from the USA, the UK, and Europe to even places like Siberia and the Seychelles. Our networks typically attract the innovators / the early adopters – who believe in Mobile technology, business and innovation despite the doom-and-gloom of the dot com industry around them.

Why don't a thousand flowers (entrepreneurs) bloom in the Mobile data industry? How can we foster innovation within the Mobile data industry? And how can we combine multiple services, bodies of knowledge and trends to create a new commercially successful Mobile data service? These questions provide the central discussion themes for OpenGardens. Our view is optimistic. We believe that there are opportunities in the existing and proposed Mobile ecosystem. We used 'Air Graffiti' or splash messaging as an illustrative application. Its principles are valid – in the sense that - given a 'set of simple building blocks' there are no limits to what can be achieved through innovative combinations of these blocks.

We also believe in an OpenWASPA (Open Wireless Applications Service Provider Association) model – a simple structure that can unleash innovation in the Mobile data industry. In that sense, OpenGardens acts as a catalyst – we don't know how you will individually use the foundations we have laid out here - but we are keen to find out through your feedback on our web site and blog on www.opengardens.net. We hope you enjoy reading this book and that it acts as a catalyst fostering new, visionary services.

Acknowledgements

We would like to thank members of our network for their practical feedback and insights especially members of the Mobile applications club and the wireless ecademy.

Ajit Jaokar would like to thank his family and Dr Peter Gray for their support. Tony Fish would like to thank Nicky, Eleanor and Emilia for their help and support.

We would especially like to thank Mike Short for his ongoing guidance and support to the book.

Special thanks go to Paul Golding for his detailed comments, Bill Volk of Bonus Mobile for his help on games models, Tom Hume from future platforms for help with developer insights and Tomi Ahonen for his insights and guidance.

Finally, we thank Andrew Darling and Maggie Baldry for the editing and proof reading.

Making the best of OpenGardens

OpenGardens can be used either as a single point of reference for principles or as a roadmap by developers in understanding future services.

We have tried to explain concepts from first principles where we can. Our vision is encapsulated in the last two chapters of the book (i.e. 'Understanding the mind of the Operator' and 'OpenGardens revisited'). The previous chapters build up to the insights in these final chapters. The early chapters explain the components that may be combined to create new services – making OpenGardens a comprehensive book.

1 Chapter One: Introduction and Scope

1.1 Insight and history

Why don't a thousand flowers - i.e. entrepreneurs - bloom in the Mobile data industry?

OpenGardens is a book for the innovator - i.e. the person with an idea who wants to create a commercially successful service within the Mobile data industry. While the Mobile data industry holds considerable promise in future, the existing ecosystem is challenging for the 'grassroots entrepreneur'. This book acts as a guide by offering a two-stage roadmap. Firstly - how to work within the existing ecosystem and then how to prepare for a more complex ecosystem of the future. We believe that without an 'innovative component', new services will not succeed in this industry (especially when they are not backed by strong brands, extensive funding etc). Thus, innovation is necessary – both for the new entrant but also for the industry as a whole to thrive. Further, we believe that opportunities exist 'on the fringe' by combining one or more elements to create a new service.

The typical reader of OpenGardens is a 'person with an idea' – who we will call the 'innovator' or the 'developer'. From the perspective of the large providers such as the Mobile operator, this person is a 'third party developer' or 'external developer'. Typically, the developer interfaces with the Mobile operator through the Mobile operator's third party developer program.

We start our discussion with an insight from that great management guru – Scott Adams – creator of the cartoon strip 'Dilbert'. [1] At the height of the dot com boom venture capitalists funded many a dubious business plan. 'Wally' (a character in the cartoon) presented a business plan to a group of venture capitalists that involved the creation of a rival 'word processor' on the Microsoft Windows platform. Even Wally – with all his skills - couldn't get that plan funded. Yet, we see many a business plan with this 'Dilbertesque' theme where you could substitute 'Microsoft' with your (friendly?) neighborhood Mobile or cellular operator. Extending this analogy to the Mobile applications development arena, many ideas depend on

---

[1] www.dilbert.com

endorsement from a major partner such as a Mobile operator - yet the idea itself has little added value for the major partner and thus rarely gets off the ground. Today, we don't have an army of software developers/innovators working profitably (with an emphasis on **profitability**) on next generation Mobile Telecoms applications. As a result of this, the industry can only achieve a fraction of its true potential.

In contrast, looking at it from a Mobile operator perspective, much work needs to be done. Recently, the chief operating officer at a UK based 3G i.e. third generation Mobile operator said

'People don't want open access, that's not what our customers tell us they want,' 'Anyone in their right mind who tries to do anything on the Internet with a screen that size has to be nuts.' [2]

As we said, much work needs to be done!

But there are other voices from the Mobile operator community that foster a different view – much closer to our own. This, from a white paper from Sprint by Robert Hammond and Hugh Fletcher [3]

Some carriers will balk at the idea of exposing their networks with a standards-based gateway such as Parlay and ParlayX. These carriers prefer a "walled garden" approach, believing the need to protect their installed customer base outweighs the potential revenue from new customers. However, the exact opposite is true. Metcalfe's Law, which states that the value of the network rises by the square of the connections to the network, proves true time and time again. For example, when North America finally decided SMS was not a fad and agreed to intercarrier message delivery, SMS traffic grew quickly. The ability to communicate easily creates scale and that is what drives profits for carriers. Carriers must differentiate on services, execution, quality and innovation, not on artificial barriers.

---

[2]   Source: Source: http://www.newmediazero.com/lo-fi/story.asp?id=249312
[3]   source: Telecommunications as an IT service – Robert Hammond and Hugh Fletcher – Sprint (www.sprint.com)

While an open ecosystem is ideal from a developer's perspective, change will occur in a phased manner. As innovators, our challenge is to understand the phases and the opportunities they create for new services.

1.2 Bridging the gap

Our goal is to act as a bridge between the Mobile Telecoms world and the Applications development world - helping developers to build Mobile Telecoms applications. We would consider our efforts fruitful if we succeed in encouraging software developers to create commercially successful services for the Mobile Data industry.

Though not immediately apparent, the timing is ideal for developers to be considering Mobile Data services. In spite of the economic turmoil, 2003 saw record handset sales – with estimates ranging from 510 million to 533 million – a 25% increase on 2002. Crucially, these are 'rich' handsets that foretell the possibility of rich services. We expect that 2004 will achieve another record.

While originally designed for voice, Mobile phones are also capable of running applications handling data – giving rise to a whole new industry i.e. – The Mobile Data Industry. The Mobile Data Industry is all around you - walk into most public places and you can't fail to be distracted by the chimes of ringtones. In addition, there are the silent devices – such as PDA (personal digital assistants), RIM blackberry [4] etc. and they are also busy sending and receiving data.

Over the last few years, the growth of Mobile Data (as opposed to voice) has shown dramatic increases, now accounting for up to 25% of a Mobile operators revenues. The industry is still in its nascent phase. It is still an emerging industry - characterised by relatively simple consumer applications like downloading ringtones. Today, services are 'entertainment led' as opposed to 'utility led' but increasingly we are seeing more utility led Mobile services as the market matures.

---

[4] http://www.blackberry.com/

1.3 Relationship between Openness and innovation

The Mobile data industry is a nascent industry and in this phase, has much to learn from outside its own domain.

Let us consider two related questions

a) What is the impact of **not** fostering grassroots innovators? In that case, all new ideas are expected to come from the existing players. Is this scenario feasible? Is it conducive for the overall growth of the industry? and

b) Is the innovator – especially from outside the industry – necessary for it's growth

The Mobile data industry currently displays 'convergence of thinking'. For example – even when there is more than one operator in a country, most personnel in the newer operators come from the older, established monopolistic operator. Thus, if new ideas are not introduced from within and outside the industry, the industry will never attain its full potential. Historically, the best way to foster new ideas is through a 'startup culture' i.e. encourage the grassroots developer/innovator. In emerging fields, new ideas often come from outside the industry – for example the work of Josef Gregor Mendel – a priest – which made a difference to the field of genetics in it's early stages through his observations on the diversity in pea plants.

Another way to look at this problem is to study ecology. The bottom of the food chain (the plankton – if you take a marine biology example) is expected to feed the 'whales' at the top of the food chain). But, the food chain can only work if the 'plankton' i.e. developers survive. By the time the meagre revenue trickles down to the developers, they do not have a viable business.

Viewed from this perspective, OpenGardens is a cause i.e. an attempt to make a difference - It is a cause we deeply believe in - within an industry we understand at the grassroots. However, our real life experience lends us to add caution. There is little point in experiencing another 'dot com'! Hence, the innovator's vision needs to be tempered by caution.

The fear of a dot com is not the only reason the Mobile data industry resists unfettered innovation. The telecoms industry

does not like to 'rock the boat' so to speak – and for good reasons. The industry has long followed standardization because by definition, every device (previously only a voice phone) must be able to speak to any other device anywhere in the world!
We simply cannot have some phones that cannot make calls to other phones i.e. every device must 'speak' to every other device. There is a lot at stake if the system fails – not only to individual users but also to security, emergency services, governments etc.

Further, the actual standardization work is complex – and is influenced by government regulators and managed through a set of committees – all of which takes time. Thus, standardization could be viewed as an antithesis to innovation. The alternative is to have companies create de-facto standards – which has its own problems – seen most commonly in the IT (Information technology) industry. In the IT industry, even when there are standards (either open or closed), these are relatively simpler than telecoms standards because of two reasons. Firstly software standards are relatively easier than hardware standards i.e. the physical component is lower. Secondly the risk of failure in a PC/Internet environment is not catastrophic.

Even in a networked environment – either a local area network or the Internet – the failure of a given node does not have a sweeping impact on the whole network. That is not the case with a telecoms environment (in this case a node being defined as telecoms infrastructure equipment rather than a specific phone). Failure could arise from many sources – software failure, hardware failure, incompatible systems, failure under stress testing, communications failure, failure during upgrades etc. The impact of all these factors is amplified in a telecoms environment. Hence, the natural approach of the telecoms engineer is to 'err on the side of caution' rather than 'lean towards experimentation'. This approach also impacts lead times – typically technological roadmaps are announced three years into the future and vendors plan up to eighteen months in the future with long implementation times for a specific product (in months).

This level of compatibility, connectivity and quality of service has been historically unsurpassed until the Internet came along!
Now the picture has got much more complex with IP (Internet protocol) devices also seeking to connect to the telecoms/Mobile

network. Hence a merging of words and the 'raison d'être' for this book. With IP come new developers and a new mindset. The Mobile operator is swamped by developers – some of whom even want the Mobile operator to 'fund their great idea' – often no more than a concept on paper.

1.4 So what does 'OpenGardens' mean?

In this book, we discuss our vision, which we call 'OpenGardens'. This is not an anti operator stance. Rather, we are looking at the same problem from a different perspective. The title 'OpenGardens' evokes a variety of responses – ranging from the open source evangelist who gets misty-eyed thinking of 'Linux on every Mobile device' to the Mobile operator who insists – 'There are no walled gardens!'

'Openness' itself can mean many things
   a) **Openness of access for the customer (i.e. the ability to access any content from their Mobile device)**
   b) **Openness of platforms (for example a level playing field for third party applications as compared to the provider's applications) or**
   c) **'Open source' as defined by** http://www.opensource.org/docs/definition_plain.php

Open source is likely to be used in an OpenGardens application. Similarly, the ability of customers to access the Internet from their Mobile device is a favorable condition for the OpenGardens environment. However, our definition of 'OpenGardens' refers mainly to 'openness of platforms'.

By 'platform', in the Mobile applications context, we mean the Mobile operator's infrastructure. An OpenGardens ecosystem is a holistic/inclusive environment, which could foster the creation of next generation utility led (as opposed to existing entertainment led) Mobile applications. These applications could often span multiple technologies/concepts and are created by 'assembling together' a number of existing applications. We call an application built on these principles an 'OpenGardens' application.

Technologically, in its ultimate form, this approach can be viewed as 'API enabling' a Telecoms network. API (Applications Programming Interface) is the software that enables service provision by the Mobile operator. The external application can make a software call via the published API, thereby creating a 'plug and play' ecosystem. The API model is also called by other names such as 'networked model', 'Bazaar model' or 'web services model'.

However, Open APIs are a subset of OpenGardens and are not synonymous with OpenGardens. We do not foresee a 'big bang' opening of the telecoms platform. The challenge is to work within the existing environment and understand the phased opening up of the platform and the opportunities that this will afford us.

Whilst open APIs are ideal for OpenGardens applications even if platforms aren't always open, we can still create applications, that span multiple elements (albeit not easily). Such applications are possible even today. They blur the boundaries between the Internet and the Mobile Internet. They require a different mindset and a true understanding of 'Mobility'. Finally, Openness – has two facets – commercial and technical. Hence, this book covers both.

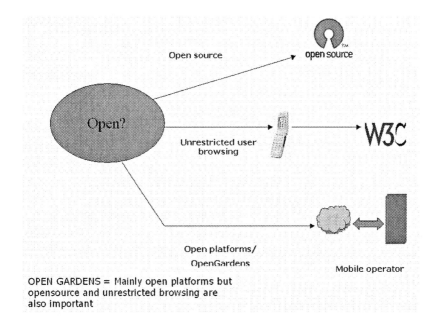

Open source

open source

Open?

Unrestricted user
browsing

W3C

Open platforms/
OpenGardens

Mobile operator

OPEN GARDENS = Mainly open platforms but
opensource and unrestricted browsing are
also important

For reasons we discuss in detail below, we focus on 'Innovation'. Innovation can often arise from simple components. Consider for a moment, the idea of 'integrating' existing applications to create a new application. This approach is not new. In fact, it is historically consistent with the methods of Leonardo Da Vinci. Leonardo was a practical innovator i.e. someone who looked at developing visionary but useful applications. His technique was simple. First he understood each part of the existing machine in detail and then sought to combine the parts thereby creating a new invention or making improvements to existing inventions. If he had merely confined himself to dismantling and assembling objects and concepts, he would be good mechanic but not a good scientist. To the process of dismantling and assembling – he added other methods such as an appreciation of interconnectedness, inspiration from arts, testing a hypothesis with practical observations and so on. These techniques give us another clue about the origins of the 'assembling together' approach in the philosophy of OpenGardens. First, understand the existing services and then create a new service by combining together a number of existing services.

At the moment, in relation to the Internet, the Mobile data industry is perceived to be slow and monolithic. This is a myth as we shall show later in the book. In the general media, there is no shortage of scepticism about the Mobile data industry considering the experience of WAP (Wireless Application Protocol). It is also easy for the rest of the industry especially consumer/marketing type experts to say from this 'We told you so'. However most people don't realise how complex the Telecoms environment/infrastructure really is and the regulatory constraints it operates under. By highlighting both the opportunities and the pitfalls, we hope that this book will act as a catalyst to stimulate your imagination and to create new ideas and concepts not yet conceived. After reading this book, you should be able to understand the Mobile data industry and create the intellectual foundation for a potential business through your own innovative ideas.

1.5 OpenGardens compared to walled gardens

As the industry matures, there is more scope for innovative, complex application that span more than one concept/discipline (what we refer to here as the **'OpenGardens application'**). These applications fulfil some customer need in contrast to being merely a source of entertainment. Such applications require the adoption of a more holistic, convergent view of the world blurring the boundaries between the Internet and the Mobile Internet.

For Applications developers, the Internet has always had a relatively restriction free development ecosystem. The Mobile Internet is closely related to the Internet itself and it's tempting for developers to apply the same paradigms to Mobile applications development. However, on first glance, the contrasts between the two environments are striking - the Internet with its free, open source movement and the Mobile Internet, with its supposed 'walled gardens'. A **'walled garden'** is a mechanism to restrict the user to a defined environment i.e. forcing them by some means to stay within the confines of a digital space. This restriction, often defined by a single company, is a means of exercising control and supposedly maximising revenue. On the Internet, the relative lack of walls (i.e. barriers) creates an ecosystem of   'May a thousand flowers (entrepreneurs) bloom'. Sadly, the lone entrepreneur (the proverbial person in a garage) has not fared as well on the Mobile Internet.

Why don't a thousand flowers bloom on the Mobile Internet?

OR
How can 'a person with an idea' create useful, commercial, innovative applications in the Mobile Data industry?

This question lies at the heart of the OpenGardens philosophy and is the focus of our book. Ask most developers what the real issue is – and their response is to often blame the Mobile operator who creates the supposed 'walled gardens'. But... Are 'walled gardens' the only problem?

1.6 Why focus an innovative component?

We believe that if you are a grassroots developer/garage developer in the Mobile data industry – your service/application must include an innovative component. Further, competitive advantage can often be availed through opportunities at the 'fringe' i.e. combining more than one elements. To understand this argument – we must consider three aspects. Firstly 'what's a garage'? Secondly 'the hurdles that an innovator needs to overcome in terms of revenue models' and finally 'Why do opportunities exist on the fringe for new entrants who do not have a competitive position in terms of brand, exclusive content etc?'

When used in this context, a garage includes an entity or individual who **has an unfavourable negotiating position with the Mobile operator**. It's easier to explain the above in context of the converse i.e. entities who **have** a favourable position with respect to Mobile operators. These include – well known brands like 'coke', large communities like 'Friends[5] Reunited', unique content like 'CNN' and anyone who has a large amount of money to spend!. Unless they have a unique proposition, practically everyone else is in the 'garage' (even if they are venture funded) because they are approaching the operator from an unfavourable negotiating position. Thus, it's a large garage!

---

[5] http://www.friendsreunited.com/

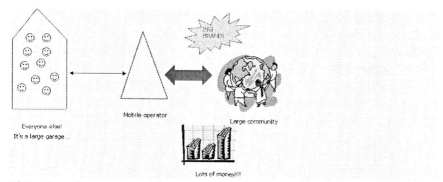

Everyone else!
It's a large garage ...

Mobile operator

BIG BRAND!

Large community

Lots of money!!!

This situation means that you need to differentiate. You cannot take on the existing players head-on. This is a large market. It's a market that will not thrive unless effective partnerships are formed. Thus, there is a need for new players and new partners in the value chain. However, starting 'in the garage' as we define it above implies your strategy has to be different from that adopted by entities that are 'not in the garage'.

Let us use the example of Mobile games here. This sector of the industry has shown promise recently. There are already players in this sector who are creating content i.e. games which their customers want. We see a critical mass of handsets in the marketplace, billing is possible and aggregator models already exist. However, developing a 'me too' game in the industry at this stage is a recipe for failure because the market is already saturated with many Mobile games.

The numbers present a sobering picture. How many game downloads would it need to attain a profit? The Mobile game industry is similar to the PC or console game market. Traditionally sales are heavily skewed to the top ten titles, and in many cases the top five take a huge portion of the entire market. Most titles lose money. Net margins for Mobile operator and aggregator distribution models (for example using aggregators like Cellmania[6], Mforma[7] etc) run from 60% to 75% in the USA. In other markets, it could be worse (for example India), or better (for example Japan and Korea). In the USA, a game typically retails at $3.99 (for an 'over the air/OTA') delivery. Product development costs for a single player game title could range up to $40K including handset testing and porting (but that could get a lot cheaper if developed overseas).

---

[6] http://www.cellmania.com/
[7] http://www.mforma.com/

Assuming a net of $2 after all costs, we need sales of 20,000 units just to break even. This gets even worse because many successful titles depend on a brand – for example a 'Harry Potter' game. Brands have their own costs leaving even less for the developer.

Therefore, we believe that most existing titles are not making money – so new entrants have an even more uphill task.

But there are opportunities in this sector today because variants in the Mobile gaming space are possible – for example 3D games, networked (i.e. multiplayer games), and games for women to name just three. In these variants, there is some scope for innovation and new players but note that each of these examples must broaden the scope of the basic offering along a new dimension i.e. a simple Mobile game. For example – combining the basic game with new technology (3D), combining the game with a community (multiplayer games), and combining the game with a demographic sub sector (women).

OpenGardens hopes to facilitate this type of thinking by acting as an 'ideas pot' where you can understand the building blocks that could then be combined in a unique way to create a new service.

This leads to the third question – why could opportunities exist on the fringe? Apart from the innovative potential of 'combining existing elements' discussed above – the competitive landscape is favourable to the new entrant at the fringe. Consider a simple example such as 'creating a new Mobile information service that would give a familiar 'look and feel' to the overseas travelers. For example – a visitor from Japan coming to London could find the best Japanese restaurants through this service in the Japanese language. This simple service spans many elements but does not come under the domain of any one player (i.e. the Japanese Mobile operators, the UK Mobile operators, airlines and so on). However, it could be marketed to travel agencies with relatively low cost. Hence a potential opportunity arises because travel agencies are relatively easy to target in marketing terms.

1.7 A word of caution

The discussion in the previous section is not the only source of concern for the innovator. The Telecoms environment and its slow moving, complex decision-making processes perplex most developers. This is more so with a pure marketing driven idea that is likely to be a concept only (no code) and is not defensible (i.e. can be easily copied). Hence, a word of caution is needed before we delve deeper into the subject. When we first discussed the idea of this book with a well-known Telecoms industry veteran, he said that the book should come with a 'Health warning'. Keep away! Too many have gone down this path with visions of gold rush 'la Klondike' but have died a snowy death. If you are like us, you will keep going anyway but let's explain what his view is – and why it deserves careful thought before you do indeed proceed.

Here is a brutally frank paraphrased view point, coming from a Telecoms industry insider -

"Telecoms has always been a profitable business for the operator and this extends to the Mobile network. In it's early stage, and even today, Mobile is still more about voice than about data.

In most countries, telecoms/Mobile have been a monopoly – with simple services, which did not need to be marketed actively.

In spite of the setbacks from the dot-com mania and the 'momentary lapse of reason' with 3G licenses, voice is still popular – both on the landline and the Mobile environment and will continue to be so.

In the dot com world, there were calls to 'subsidize Telecoms' for the 'greater good'. No industry can afford to do that and in retrospect, the proponents of 'free' are assigned to the digital graveyard of the dot-com industry. With the rise of the Internet, the fixed line Telecoms operators were relegated to being 'data pipes'. (In the sense that the big stars of the Internet were Yahoo, Google, Amazon etc. who sold services over a network operated by the fixed line Telecoms operator. These businesses were far more profitable but the fixed line Telecoms operator ended up being the 'data pipe' and seeing little revenue.)

When Mobile data came along, initially, there were hopes of 'we missed the Internet, but we will NOT miss this one!' There was, and still is, a tendency to 'go it alone' without partners. It is debatable as to how much value partners add - unless they happen to be well known brands like Warner and Disney. Most ideas are not new and most concepts presented to Mobile operators are merely at the concept/planning stage.

While this is changing, the fact remains that barring SMS, Mobile data revenues are very low at the moment (early 2004).

Does this mean that the Mobile operators do not need partners at all?
No...
There are three reasons why they do
a) Firstly, although Data is taking longer to become mainstream, no one denies that it's going to be a huge source of revenue for all concerned
b) 'You never know who walks in through the door' – May be there is a killer app out there?
c) Related to (b) – competitive intelligence – someone else may deploy a killer application first"

Although this frank viewpoint comes from the limited perspective of a Telecoms Operator, it is well worth taking it into consideration whilst you are developing your application. You should also contrast this with the 'OpenGardens' approach, which deals with creating new, composite applications from existing applications, industries.

1.8 Definitions and core themes

As we begin our journey, let's emphasise some of the core themes we are going to use in this book. As you probably don't know us or have heard us speak – you may want to read these definitions to understand our interpretation of these terms.

Mobile Internet – 'Mobile IP data service'. It is not 'Internet on the Mobile device' since mobility also includes other elements such as 'messaging' i.e. non-browsing modes of access.

OpenGardens – When we refer to 'OpenGardens', we are referring to 'Open platforms' and next generation holistic, inclusive applications which span multiple technologies/concepts often created by combining existing services.

Walled Garden - an attempt by an organisation or entity to restrict or direct the user in some way and control the economic model.

Person with an idea - refers here interchangeably to developer, entrepreneur, and startup, visionary inside a company or even to a visionary outside the Mobile data industry itself who is building a new application.

Innovation – refers to both technical and business process innovation.
In addition, OpenGardens is not a visionary book – in the sense that it speaks of the 'here and now'. Our main goal is to correctly (and independently) describe the existing landscape and define the services on the horizon, which could be profitably exploited. **In this sense, OpenGardens is a 'tactical' book.** As the industry itself evolves, this horizon will shift i.e. future versions of this book will focus on different services.

Geography: While both the authors are based in the UK but the principles outlined here apply to most markets.

Mobile vs. wireless: In Europe, the commonly used phrase for Telecoms data applications is 'Mobile'. In USA, it is 'wireless' or 'cellular'. In this book, 'Wireless' simply implies connection without wires. Mobility or 'Mobile' on the other hand describes a whole new class of applications which permit us to interact and transact seamlessly when the user is on the move 'anywhere, anytime'. Hence, we use the term 'Mobile' independent of technology i.e. 3G, wireless LANs etc.

Service vs. application We use these two terms interchangeably. The IT industry is more familiar with the term 'application' whereas the telecoms industry uses the term 'service'. In the purest sense, a service includes an application i.e. it includes other components like support whereas the application implies the code only.

Are you a pipe-hugger? Do you wake up each morning and think about how wonderful your gas, electricity, water suppliers are? No? Then, why do we compare 3G/Broadband/Wifi? Pipes all … In practice, they are irrelevant in context of a larger, integrated world as long as they continue to shift increasingly larger bandwidths.

1.9 Geographical differences

The Mobile data industry has evolved at differing speeds in various parts of the world. Japan has an advanced usage of Mobile devices. South Korea shows increasing trends towards digital convergence with a high adoption of broadband. Europe, while being a large market as a whole, is fragmented in terms of language and culture. North America has a high uptake of BREW[8] from Qualcomm[9] and so on.

We expect that these local factors will continue to prevail i.e. the market as a whole will have global similarities (for instance adoption by youth, location based services, digital convergence etc) but at the same time will have local differences. We outline some geographical variations below.

1.9.1 Japan

Japan has three operators - NTT DoCoMo[10], KDDI[11] and J-Phone[12] with NTT DoCoMo (i-mode) being the market leader. Both entertainment based applications and functional applications have long been possible such as paying for services, finding the nearest restaurant etc. Japan has a culture focussed on visual content – for example Manga (comics). Similar trends towards visual content are seen in South Korea and China and are reflected in phone personalization.
Important content players include Cybird (http://www.cybird.co.jp/english/), and Xing (Japan's largest supplier of ringtones) http://www.xing.co.jp/english/index.html

1.9.2 South Korea

Like Japan, South Korea has a visual culture and a population that loves new technology. South Korea has a high broadband penetration and a high usage of online networked games (multiplayer games). The technology industry in South Korea has strong government support with an emphasis to create global domination in multiplayer games and home networking. South Korea also has a high adoption of Brew from Qualcomm.

---

[8] http://brew.qualcomm.com/brew/en/
[9] www.qualcomm.com
[10] www.nttdocomo.com
[11] http://www.kddi.com/english/index.html
[12] http://www.vodafone.jp/scripts/english/top.jsp

There are three operators in the Korean markets: SK Telecom[13], KTF[14] and LG Telecom[15] with SK Telecom being the market leader. There is high adoption of video on demand in South Korea and content is dominated by video. The "June" portal from SK telecom provides video oriented services closely tied to TV shows. South Korea also has strong device manufacturer brands such as Samsung (www.samsung.com) and successful Mobile entertainment companies like Com2Us (http://www.com2us.com/english/index_e.asp)

1.9.3 China

China telecom is the world's largest telecom operator and the Mobile phone culture is growing at a rapid pace in China. China does not have a high penetration of other devices – such as handheld game players like Nintendo game boy – making the Mobile device all the more important. The Mobile phone culture in China is also influenced by South Korea – especially by the South Korean TV dramas which are exported to China. Similar to South Korea, we see an incidence of Online gaming culture

The size and importance of the Chinese market leads to some unique characteristics for example the Chinese government is proposing its own 3G standard TD-SCDMA.

1.9.4 Europe and the European Union

Collectively, Europe has a large and affluent population. However, Europe is culturally and linguistically fragmented especially with the joining of ten new Eastern European countries in May 2004 into the European Union. Technically, there is one standard GSM (Global System for Mobile communications) with a strong uptake of SMS (text messaging).

There is a high level of phone penetration in Europe averaging 80% in Western Europe. Europe has strong brands like Nokia. Some countries – especially Scandinavian countries had an early lead with Mobile. However, 3G deployments has been slow overall in Europe and the early adopter mantle has been taken up by countries like South Korea.

---

[13] http://www.sktelecom.com/
[14] http://www.ktf.com/
[15] http://www.lgtelecom.com/eng/index.jsp

1.9.5 USA and Canada

Although USA lags Europe in Mobile phone penetration, it is an important market due to its affluence and dominance of media. USA has six major Mobile operators with Verizon Wireless the largest. USA is fragmented in the adoption of technology between. Canada is similar to the USA with Rogers AT&T and Bell wireless being the main Mobile operators.

1.10 Mobility and Digital convergence

In the context of the wider digitization and digital convergence taking place around us, the scope of this book is limited to its coverage on mobility.

The book does not cover technologies such as RFID (Radio frequency identification), Satellite, Digital television etc. However, keeping our philosophy of 'opportunities at the fringe' – these technologies are a source of innovation. As more content gets digitised, greater bandwidth is possible and the quality of the transmission improves – the possibilities of new, innovative applications increase.

The deployment of IPv6 increases the Internet address space to 3, 4 x 1038 address many of which will be taken up by new devices. The impact of this change will mean many more devices can interact with others. Although not within the scope
of this book, clearly there are fantastic opportunities in this space!

2 Chapter Two: Basic Concepts

2.1 People

We start our discussion with people rather than technology.

The Mobile data industry (which commenced around 1996) is closely related to the rise of the Internet. In the dying phases of the dot com era – the Mobile data industry became the last straw which many attempted to grasp (much to their dismay). Due to this genesis, the Mobile data industry attracts a diverse range of players. This history of the Mobile data industry is reflected in the people that you are likely to encounter. These include -

- **The refugee from the dot com industry** – "I lost a (paper) million in the dot com boom but I am going to make my next million here!"

- **The maverick** – "Hmm – what have we here?"

- **The innovator** – "I want to be an entrepreneur. I have a single idea / concept which I want to test out"

- **The serial entrepreneur** – The innovator with money

- **The Venture Capitalist** – Who sees it as a new sector to make money but is wary of the stigma of dot com.

- **The Telecoms engineer** – He has been in the Telecoms industry for many years (and yes, it's almost always a 'he'). Knows it well from an Operator/Telecoms point of view. For the first time, he is exposed to the consumer market directly.

- **The advertising/marketing executive** – Aware of brands, Internet, marketing etc. but clueless when it comes to Mobile. Thinks all problems could be solved only if there were the right marketing/branding focus.

- **The media/content owner** – Who owns/manages rights to content. Approaches the industry with a view to monetize that content. Probably already has a similar venture on the web.

- **The web designer /developer** – Already familiar with the Internet. Seeks to extend his knowledge to the Mobile Internet

- **The community builder** – "I have built a web based community and I think I can extend that to the Mobile sphere OR I want to create a Mobile community"

- **The enthusiast** – Just interested.

- **The retailer** – Got the first taste of success with 'Mobile marketing'. Looking for more.

- **The infrastructure manufacturer** – Telecoms equipment manufacturer. Creates 'Telco grade' applications/ software i.e. software applications meant to be installed at the Mobile operator's premises.

- **The executive** – Who likes to think they understand this space

- **The development manager** – Who is responsible for third party engagements

- **The Mobile portals/aggregators** – Similar to Internet portals, trying to make themselves the first port of call.

- **The device/handset manufacturer** – Works with a company like Nokia, Sony-Ericsson etc.

- **The Telecoms Operator employee** – Not an engineer but in the Telco industry. Works for BT, Vodafone, AT&T etc.

- **A 'point of interest' on the Mobile Internet** – This includes everyone from the art gallery to the corner newsagent who are often at the end of a 'Find my nearest' query.

- **The perpetual optimist** – Who hangs out with little cash but with high hopes and finally – the most important person

- **The subscriber** – Also called the customer. The person, who pays for the service by subscribing to the Mobile operator's network, pays the bill and calls when things don't work. Subscribers are increasingly becoming more than 'consumers'. In some cases, they are becoming producers of content by being active participants in the value chain.

A byproduct of this diversity is the lack of a holistic view – i.e. groups often see the world only from their point of view – sometimes missing the big picture. This crucial factor is more important than the walled gardens themselves! To rephrase, it would make a lot more sense for all parties to understand each other to create a 'bigger pie' through synergy. We will come back to this aspect again in our discussions.

2.2 The most important person – the customer

The customer is the one who pays for the service. As with any industry, the goal is to 'serve the customer' – the most important person in the list above. From the customer's perspective, she sees two entities – the device manufacturer (who makes the phones) and the Mobile operator – who provides the connection. Through the device – the customer accesses a range of services. Primarily, these are voice – since the Mobile phone is still predominantly a voice oriented (rather than data oriented) device.

A service is more than the application (software) – it could roughly be described as an 'application deployed' i.e. the core application plus support structures such as billing, customer services etc. The customer uses a service. Broadly, Mobile data as a service could be classed into a 'content' based service or a 'contact' based service.

In the existing Mobile operator-centric view of the world: -

- In the content-based service, the 'content provider' to the customer, via the Mobile operator, deploys the content (information).

- In the contact based service, two or more people communicate with each other via the network.

As we have noted, these are 'Mobile operator centric' views of the world. This view is not valid in all cases (for example two people could communicate over Bluetooth or a wireless LAN i.e. an ad-hoc network without involving the Mobile operator) but it is sufficient for our discussion at this point.

2.3   Content based view of a service

From a big picture perspective, the key difference between a web based application and a Mobile application is the presence of a Mobile operator/carrier in the transaction in most cases (i.e. the deployment of the service over an air interface). Traditionally, content flows from the content provider to a Mobile subscriber through the subscriber's Mobile operator. The Mobile operator bills the customer for the content and the content provider receives a share of this revenue.

Mobile operator revenue models are based on the concept of ARPU (Average Revenue per User). The billing entity (such as the Mobile operator) bills the customer according to usage (based on time connected, a one-off charge or the amount of data transmitted/received). That amount is reflected in the user's phone bill. The revenue thus generated can potentially be shared with the content provider/promoter. The Mobile operator's overall goal is to increase ARPU and one way of doing this is by providing more and richer functionality. Currently, voice applications drive the Mobile market – and voice is expected to dominate for the foreseeable future, but with digital/data applications being increasingly important.

The term 'content' could be used loosely as 'information' which the user/customer seeks to access via their Mobile device. In a more rigid/traditional definition, a content provider is a company which owns/aggregates specialist content, which is served to the consumer (such as a mapping company that renders a mapping application but does not own the mapping information itself). Thus, the content provider may not be the owner of content.

Further, most countries have more than one Mobile operator (for example, even in Japan where NTT DoCoMo is the most recognized, there are two other Mobile operators - KDDI and J-Phone). This means that the customer using your content may not be on the same network to which your application connects. Thus, you need to be sure that the Mobile operators have appropriate interconnect agreements signed for delivery and revenue collection if you want to access customers from more than one Mobile operator.

The Mobile content industry is an amalgamation of two distinct industries - Mobile and media/ entertainment. Until recently, these industries did not have much in common. Mobile (wireless) was a part of Telecoms and functioned as a utility. Entertainment, on the other hand, was much more customer facing. Potential market opportunities have brought these two industries together but their internal cultures do not mesh well. Hypothetically, imagine what would happen if 'Warner Brothers' tried to partner with 'Thames Water' (a UK water company) – and you get the scenario of two industries each in their own world.

The relationship between the two industries (Media/Entertainment and Mobile) starts with Entertainment providing the content that feeds the Mobile value chain. In a nutshell, in the Mobile value chain, revenue starts with the subscriber and flows to the content owner via the Mobile operator.

The Mobile value chain is as depicted below:

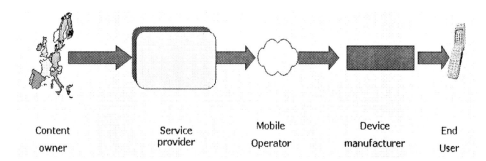

| Content owner | Service provider | Mobile Operator | Device manufacturer | End User |

Characteristics of Content owners:
- Content owners include players like Broadcasters (TV, radio etc.), News agencies, Publishers, Entertainment companies (movies, music, entertainment), Rights Owner Companies (music rights, sports rights, general showbiz agents etc.)

- Unlike the Internet, the general perception of the Mobile Internet is that content is not free.
- Own the rights and /or represent the copyright owners
- Large content owners are trying to negotiate directly with the Mobile operators
- Content owners understand the power of their brand
- They have complex models such as based on geography, mode of access
- Believe that 'Content is king'
- Are trying to negotiate 'upfront' payments for their content but are often having to settle for revenue share
- If they are strong brands, they are aware of their value to the Mobile operators

Characteristics of Content aggregators/Service providers:
- Examples: 'Vodafone Live!' From Vodafone
- Could be an Mobile operator portal or an independent portal
- Act as an aggregator in the market
- If they are a part of the value chain, they are concerned with billing
- They are customer facing
- Believe in high volumes ('Pile 'em high – sell em cheap' as one Content Provider told us)
- Work on the revenue share model but would like upfront payments where they can

Characteristics of Mobile operators:
- Example: http://www.t-Mobile.com/, http://www.verizonwireless.com/, http://www.nttdocomo.com/
- Have a direct relationship to the subscriber
- Influence the whole value chain
- Negotiate deals with the third parties

Characteristics of Device manufacturer:
- Example: Nokia, Sony Ericsson, Samsung etc.
- Can be strong brands in their own right
- Are often the first physical point of interaction with the customer
- Could enable a rich media experience through better and richer devices
- Are involved in DRM (Digital Rights Management) mechanisms on the client side

- Are trying to own a larger share of the value chain by becoming portals. For example - Club Nokia

Characteristics of End Users:
- Actual consumer of content (pays for content)
- Defines the market (demand)
- Believes the 'Customer is King'

2.4 Contact based service

A contact based service involves two or more people communicating with each other. For example by sending a text message.

Contact influences content. While the mantra of 'Content is King' is well publicised, the definition of content is often narrow. The general perception of the content creators is to follow the 'broadcast to consumer model' i.e. the content provider broadcasts while the consumer passively consumes. This model often ignores the active interaction and in many cases creation of content by the consumers themselves. Content created by the customer (or at least customised by the customer in some way) is often potentially more valuable to the customer than the content, which is simply broadcast.

We believe in contact over content – in fact we believe 'content' will be strongly influenced by 'contact' in many cases making present day content more similar to contact in future.

2.5 Ahonen's five 'M's

Another way to look at services is to consider the Five 'M's which Tomi Ahonen refers to in his book 'Services for UMTS' (Wiley – Tomi Ahonen and Joe Barrett). (UMTS is the Universal Mobile Telecommunications System)

The five 'M's are:

Movement – escaping place

Moment – expanding time

Me – extending me (personalization)

Money – expending financial resources

These can best be explained by an example – also from Tomi's books.
On a rating of 0-5, the Mobile Ringtone service fares as follows with the 5Ms

---

Movement
Rating = 5.
Because I can download a ringtone from wherever I want and it's simple

Moment
Rating = 5.
I can select the moment that is convenient to download the ringtone. If I see the ringtone promoted on TV, I can download the ringtone at that moment

Me
Rating = 5.
The ringtone is based on my tastes and dislikes and it says something about the kind of person I am

Money
Rating = 5.
I don't mind paying for the ringtone since it's quite inexpensive.

Machine
Rating = 3.
When I download the ringtone, I do so interacting only with a machine (server).

2.6 The problem with the most important 'M'

While personalization ('Me') is important – the main element of differentiation of a Mobile service is 'Movement' i.e. location. To most people, movement (also called location-based services) is intuitive and useful. It's also something we can all relate to (for example a 'find my nearest' service when we are on the move). We will discuss location based services in detail later, however, the chief problem at the moment is, location based services are not accessible at a mass-market level in most countries.

Further, location is more than 'find my nearest'. A location-based service/application must truly take location into account at all times. For example – when you are in a foreign country, the information returned could be mapped to your current location depending on your query – not your home location (for example – nearest cash machines must be nearest to your current location and not your home location). However, if you want a stock market portfolio, it must default to your 'home' portfolio and not the portfolio of the country you are traveling in. (example adapted from Services for UMTS – Tomi Ahonen)

When location based services are truly available, Mobile services will indeed take off. The challenge at the moment is to design a service when location based services are not ubiquitous (i.e. the other 'M's take more importance purely due to the non-availability of location at a mass market level)

3 Chapter three: A Strategic Perspective

3.1 Approach

As a developer/visionary, I want to develop an application that someone will pay for. If you ask a Mobile operator the question *'What applications do they need from developers?'* – they are likely to say *'They need applications which their customers want'*. Herein starts the parting of understanding – since customers **don't** know what they want when the service is new and innovative. Who would have predicted the success of EBay, Amazon, Google, the Walkman and so many other innovations! Further, the question itself is leading – it assumes that the Mobile operator is the centre of the digital universe and even further that the Mobile operator knows (i.e. can predict) the demand for future services.

Both, as we know, are not necessarily the case. To be fair, no one can predict the demand for future Mobile data services but in light of the wider digital ecosystem we describe, a Mobile operator centric view of the world is limiting. Further, the traditional metrics of segmenting customers, focus groups etc. do not work well in this case since customers do not know what they want (i.e. they cannot ask for what they currently cannot visualise).

Alternately, instead of asking the Mobile operators, should we go to each geography and try to find best practise and business models? Would it be useful to do so? It's been a few years now since i-mode has been a success in Japan – but models similar to the i-mode model have not emerged in other places. While we can learn from these countries, each country operates under a different set of parameters such as legal, technical, cultural, commercial etc. A number of analysts have already described the i-mode phenomenon in great detail (we suggest doing a Google search to get loads of free information). Merely describing models such as i-mode in detail would be 'for information only' i.e. readable but not practical for developers in other countries. Thus, applying lessons from one ecosystem to another might not be practical since we have to wait for that ecosystem to develop in a different geography and that can take years (if it happens at all).

Considering we are not attempting to predict a specific future application but rather attempting to foster an ecosystem, which would be conducive to creating a range of new applications (which none of us can predict), let's start with the existing ecosystem. Study of the existing ecosystem will highlight the problems in creation of new services at a developer level and also the potential opportunities in an emerging ecosystem. An accepted model of existing services is the 'hypermarket' model proposed by Nokia in their white paper 'Make money with 3G services' which is depicted below.

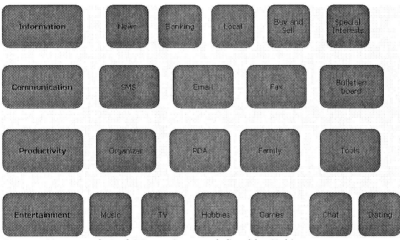

Source: Hypermarket of 3G services as defined by Nokia
Nokia, make Money with 3G services white paper referred in
Revenue Assurance, Fraud & Security in 3G Telecom Services Mark Johnson

A look at these services reveals that they are really extensions of the World Wide Web or the existing content value chain.

For example – reading mail on your PDA, executing banking transactions on your Mobile phone, listening to music on your phone and so on. Crucially, from a developer/ innovation point of view, there is little 'added value' in the sense that these services could easily be developed by existing players. For example - the company who has developed the web site or the company who develops the application (e.g. corporate email) or the entity that owns the rights to the content etc. The decks are stacked firmly in favour of the incumbent. At the most, as a software developer, one could get an assignment developing a Mobile service for a specific client but the revenue model becomes 'software/consultancy' – i.e. mostly one off assignments for clients developing software.

Contrast this with a very different type of application called 'splash messaging/air graffiti/spatial messaging'. In its simplest case, it's the ability to 'pin' digital 'post it notes' at any physical point. Suppose you were at a holiday destination and you took a picture or a video of that location. You then 'posted' that note digitally with your comments and made it accessible to your 'friends'. Many years later, one of your friends happened to come to that same place and as she walked to the venue, a message would pop up on her device with your notes, picture and comments.

We will revisit this application to illustrate the concept of OpenGardens. At this point, if we contrast the Spatial messaging application to those listed above in the Hypermarket model, we make some initial observations

a) We see that the Spatial messaging application is truly different from any applications that exist before on the Web and involves the concept of mobility at it's foundation

b) This application involves a range of elements within mobility (technical, social and commercial) for example – determination of location, ability to recognise your 'friend' through a profile, the availability and interoperability of technologies such as MMS (Multimedia messaging) which drive camera phones,

c) A spatial messaging application would be easier to build if the Telecoms network were 'API enabled' i.e. it would be possible to make 'software calls' from an application through published APIs

d) Note that it is also possible to build the spatial messaging application without open APIs in place – but as we can imagine this would be limited (i.e. not accessible to everyone) and not easy to build.

e) The application calls for 'blurring the boundaries' across many social/technical elements.

While we have hinted at the 'nirvana' ecosystem (i.e. an API enabled Telecoms platform), this is not merely a technical issue.

Technical solutions are already with us now (for example Parlay X) but commercial issues could be the 'make or break' here.

To understand the commercial issues, it's necessary to look at a bit of history – however short that is in case of the Mobile data industry.

Specifically, to understand 'OpenGardens' we need to understand the 'why and how' of it's philosophical opposite – the 'walled garden'.

3.2   An introduction to walled gardens

What is a walled garden? What are the 'bricks in the wall' i.e. the elements that make up a walled garden?

A walled garden is any mechanism for an entity (not just a Mobile operator) to restrict the user experience by confining the user to a specific region / space as defined by the entity. The rationale is - the user is served better and the service is more profitable for the provider. In an Internet/Mobile environment, this can often take the guise of restricted browsing but has other facets as we see below.

From a developer perspective, a walled garden could mean 'restricted access', i.e. - your application in some way cannot access all customers OR the provider's application has access to some features that you cannot access. These restrictions can be commercial or technical. In conversational terms, walled gardens are deemed to mean any restriction placed on users or applications, which are aimed at confining the user to a set of features controlled by the provider.

Walled gardens are not new. One of the best-known instances was the early AOL. On an extreme case, in the early days, users could not email others outside AOL! (remember this was only about eight years ago). However, the early users liked these restrictions since there was the perception of 'the big bad world out there' and AOL was deemed to be a trusted provider. As users matured, they realised that the restrictions were often a hindrance and ultimately, there are lot less restrictions now on AOL. However, even today, AOL users have a different experience of the World Wide Web. It's debatable if it's better of not – but it's certainly different.

Within the Mobile data industry, a Mobile operator has some elements that lend considerable power

a) A large customer base
b) Knowledge of the subscriber's location
c) Billing relationship to the customer and
d) Customer services and marketing reach

There are others especially on the voice side but these elements are critical for data applications. In addition, in a portal situation, the Mobile operator has the ability to control the positioning of the user on the menu, which is yet another 'Brick in the wall'.

Extending the concept even further – Mobile operators are not the only ones in the walled gardens game. Brands are often viewed as a safe bet - especially branded content. A friend uses the colourful and insightful expression 'Elephants mate with elephants'. This means the large content providers of the world may well have done deals with the large Mobile operators leaving little scope for the smaller player. There is already evidence of this with some content deals in Europe. Many in the industry blame the Mobile operators for creating a walled garden. Walled gardens restrict the potential slew of applications that could be possible if everyone were allowed to create any application and users of a system had total freedom to choose any service inside or outside that system.

The issue of walled gardens first arose with WAP (Wireless access protocol) phones, which are used to "browse" content. It arose due to a specific legal situation with a European Mobile operator who prevented users from changing the default settings on their phone. This means, users always started with a specific WAP site (i.e. home page) as directed by the Mobile operator and further they could not change the home page itself. While this model was commercially appealing to the Mobile operator and also the advertisers, it was not conducive to the small developer. A developer successfully appealed against the Mobile operator and won. In retrospect, the whole issue seems irrelevant in the case of WAP - because for various reasons, consumers never used WAP sites in large numbers.

So, do walled gardens exist?

While the 'hardcoded WAP home page' does not, there are indeed other ways to create restrictions.

A true OpenGardens ecosystem would exist if 'all applications had a level playing field'. Where 'menu positioning' seems to be the most obvious 'choke point', there are other ways to cripple applications belonging to external developers specifically if they are denied equal access to certain resources for example location information.

Taking the dispassionate view, is the Mobile operator a 'friend or a foe' to the external applications developer? (Here 'external' refers to a developer who is not associated with the Mobile operator i.e. third party developers). In our view, it's neither. A Mobile operator exists in a wider ecosystem and will continue to exist (and dominate) that ecosystem. Nor will the Mobile operator be insignificant in the sense that the Mobile operator will **not** be relegated to the ranks of a mere 'data pipe' as occurred in the Internet world. It is wishful thinking to hope that Mobile operators are suddenly going to change. They are not.

In a free market economy, it would be unheard of for an entity (i.e. Mobile operators) to give up their core asset and somehow create a level playing field for others when it's not in their interest to do so. This is not likely to happen. The only way things are likely to change is that if all players believe it's in their best interests to make this happen. There are many initiatives which approach this problem from a technical perspective but the key issue is commercial and legal/statutory not technical. Assuming that there is no spectacular legislation in the pipeline (and we sincerely hope there is not since governments have 'influenced' the industry already with the tax on 3G licenses), the industry has to be its best advocate.

Certainly, Mobile operators and developers have a different mindset. One person (working for a Mobile operator) referred to a quaint phrase 'revenue leak' – and no prizes for guessing who they think the revenue is 'leaking to'.

While most Mobile operators genuinely don't know how to tackle the brave new world of applications development, there are other factors at play here. Some Mobile operators are concerned about pricing a service 'wrongly' in their view. Their fear is – the service will be popular but not profitable. Hence, a further reluctance to relinquish control since a global 'free for all' access tends to reduce prices. Finally, we should not mistake the present scenario as a barometer for the future. Existing applications are **'broadcast content driven'** whereas future applications are likely to be **'utility content/contact driven'**. We examine this issue from an innovation perspective in this section below.

In addition, there are some issues which are controlled by neither the Mobile operator nor the developer – such as legal and statutory guidelines that we discuss later. Thanks to a certain measure of control, the Mobile data industry has been spared the worst excesses of the Internet such as SPAM, copyright violation, privacy violation etc. As the industry matures and adopts more guidelines – the question is that of 'intent'. Is control a sign of maturity or is it a smokescreen? Clearly, the Mobile operator cannot flow against the rising tide of the Internet and the global flow of information i.e. the Internet is the driver to mobility and not the other way round. The Mobile operator who does not embrace the wider Mobile data world is likely to be a minor player and become less profitable but will possibly survive none the less.

Coming back to AOL, it appears that walled gardens do not stand the test of time and become irrelevant as the medium matures. We believe this will happen in the Mobile data industry, i.e. it is not possible to predict in advance as to what content/applications a user may want. Not forgetting, of course, that user preferences change over time.

3.3 An introduction to OpenGardens

Our vision of OpenGardens can be summed up in the following ideas:

a)   We believe 'a level playing field' which encourages the grassroots developer is essential for the overall prosperity of the industry. This implies 'openness of platforms' and 'openness of access' which we defined at the outset

b)   An innovative component is necessary for garage entrepreneurs to make a difference because they lack the backing of big brands, extensive funding etc.

c)   Opportunities lie 'at the fringe' i.e. the intersection of the business/technical elements that comprise the Mobile data industry.

d)   For reasons we outline in detail later, the open 'free for all' culture of the Internet does not translate directly on to the Mobile Internet.

e)   Hence, rather than a 'big bang' opening – we are witnessing a phased opening of the gardens

f)   Ultimately, the vision of Open Gardens lies in Open APIs and the emergence of a 'plug and play' ecosystem (also called the networked model, Bazaar model, web services model). Indeed the vision of a 'plug and play' ecosystem is so vast that it affects all facets of Mobile applications development. Hence, it requires extensive understanding of a number of strategic, business and technical aspects of Mobile applications development. The model is not radical or futuristic – companies like IBM (www.ibm.com) are speaking of 'On demand economy' even today. The model can also be seen with the rise of ay and Amazon. The provider (for example Amazon) becomes the hub (and not a pipe!) in the marketplace where small providers (sellers of services) transact business profitably and easily. For this to happen successfully, we need both technical feasibility and commercial viability (for both the small provider of services and the company who manages the platform).

g)   However, Open APIs need to be looked at in the wider context i.e. the viewing Open Gardens in terms of Open APIs is one dimensional. It misses other opportunities like 'off portal revenues' that we discuss later.

h)   We believe in the OpenWaspa model that we describe below in detail.

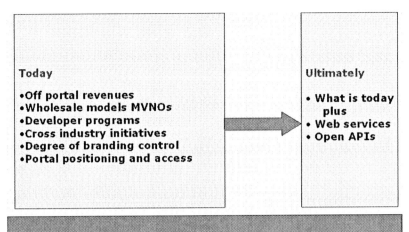

Today

- Off portal revenues
- Wholesale models MVNOs
- Developer programs
- Cross industry initiatives
- Degree of branding control
- Portal positioning and access

Ultimately

- What is today plus
- Web services
- Open APIs

OpenGardens – A phased opening. Opportunities and strategies

In the next section, we discuss the commercial elements from the perspective of creating an OpenGardens ecosystem. We then revisit the OpenGardens model itself armed with our understanding of both commercial and technical aspects. In the remainder of this section, we look at OpenGardens from a 'historical' perspective. 'Historical' meaning everything that came before the Mobile data industry. There are lessons to be learnt from studying emerging industries since we believe that the Mobile data industry is following the well-trodden path to maturity taken by other industries that came before it. If we accept for a moment, that the Mobile Data industry is a part of a larger commercial picture like any other industry, then historical insights should apply to the Mobile Data industry. Specifically, we look at aspects such as the impact of operating in an emerging, industry and the 'Innovator's Dilemma'.

3.4 The Innovator's dilemma

The Mobile data industry is an emerging industry. The evolution of emergent industries is well documented especially through the work of Dr Porter in his books 'Competitive strategy' and 'Competitive advantage'. According to Dr Porter in 'Competitive Strategy': -

"Emerging industries are newly formed or reformed industries that have been created by technological inventions, shift in relative cost relationships, emergence

of new consumer needs or other economic or sociological
changes that elevate a new product or a service to the
level of a potentially viable business opportunity"

Emerging industries have no rules since rules are being created as the industry evolves (in all spheres such as technology, commercial, legal etc.). For example – we see technological uncertainty (the deployment of 3G), strategic uncertainty (poor knowledge of competitors), difficulty in forecasting demand etc. The Mobile data industry is also a fragmented industry. According to Porter, characteristics of a fragmented industry include industry entry and exit barriers, cost issues, diversity of market needs, possibilities for product differentiation, government regulations, conditions for economies of scale and other advantages of size. Fragmentation can be overcome using strategies such as specialization, increasing the value added components, cost reduction and backward integration.

Creating revenue from services in this stage of the industry is a tricky proposition. However, it is not completely new because other industries have evolved through the same phase. Two existing studies are applicable to this problem – Firstly 'Crossing the chasm' – by Geoffrey Moore and secondly the 'Innovator's dilemma' by Clayton Christensen.

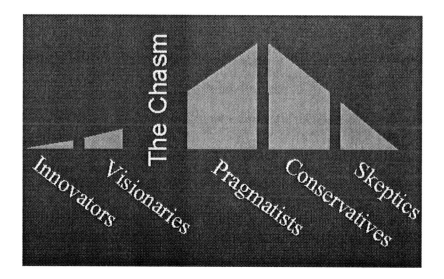

Moore divides customers by their inclination to adopt new technologies. On one hand are the innovators and visionaries – collectively called as the 'early adopters'. On the other hand – we have the pragmatists, conservatives and the skeptics. Between these two classes lies the 'chasm' which applications/products must cross to become mainstream.

Related to crossing the chasm is the work by Clayton Christensen in the 'Innovator's dilemma' (Harvard business school press). In 'The Innovator's dilemma' Clayton Christensen talks of the impact of a disruptive technology and the reaction of incumbents to disruptive technologies.

Christensen describes sustaining technologies as those improving the performance of established products along the dimensions of performance that mainstream customers in major markets have historically valued. In contrast, disruptive technologies lead to *worse* performance for established customers in the near term. However, they have features that a few fringe (and often new) customers value. The solution thus appears to be in finding a class of customers who will be willing to pay in the early adopter phase of the industry.

If the principles of the Innovator's dilemma do indeed apply to the Mobile data industry – then the innovator (i.e. the creator of the disruptive technology i.e. developers) must first find a niche to serve profitably. When we speak of a niche sector to target within the Mobile data industry, developers instinctively target 'Business' customers. This sounds logical because it's B2B (and not B2C). So, we think that we are safe! But are we? Business customers may pay the bills but the model is not easily scaleable – hence not disruptive.

Going by the history of the Mobile data industry so far, it has been mostly about communication and cosmetic applications as opposed to utility (and therefore the possibility of a disruptive application). We are living in a 'post Internet but pre mobility' world (Broadly true for most countries in mid 2004). Living in a 'pre Mobile environment' may sound strange at first but true mobility is still not a feature of applications today. We believe that the observations of the Innovator's apply more to the second phase of the Mobile data industry and not the existing first phase. In that phase, a new class of customers, as yet undefined, is likely to emerge. The crucial difference is - this class of customer will gain a direct benefit (cost or competitive) and hence will have a propensity to pay - creating a virtuous value chain.

The above discussion leads to two questions -
Firstly – what services should we consider currently which have the potential to make money? In other words – what combination of customers / services should we consider in this phase of the industry evolution (i.e. when the industry has not crossed the chasm and could still be classed as an emerging/fragmented industry?) and secondly – how do we tackle the unknown – the services which are not possible to predict at this stage.

When little is known about the future, how do we gauge demand for new services? The easiest way is to experiment on a small scale. The real question is – what to experiment with? We believe that the ultimate answer is 'web services'. But there is an evolutionary phase where it will be necessary to identify opportunities for innovators. Further, opportunities are likely to be found 'at the fringe' i.e. by combining more than one elements to create true innovation, competitive advantage and profitability.

3.5 Services – creation and deployment

In the previous sections, we discussed the need to identify services that an innovator could develop profitably in the current phase of the industry (i.e. when the industry has not crossed the chasm and could still be classed as an emerging/fragmented industry). We have also seen that in many ways, the deck is stacked against the innovator. It is our belief that the best opportunities lie at the cusp i.e. by combining multiple elements to create a new service.

Before we discuss this philosophy, it is essential to study the basic principles of Mobile services in general. In the last chapter, we will then build upon this knowledge to describe areas where we could create new services profitably.

3.5.1 Introduction to services

In a traditional IT world, we talk of applications. In a telecoms world, the commonly used phrase is 'services'. A 'service' is roughly an 'application deployed'. It includes not only the application code itself but also other ancillary functions like support, billing, customer case, quality of service etc. Understanding the service concept makes a huge difference to the success of the venture. For example - in many cases, the Mobile operator is the first port of call for customer support calls. This implies that the operator is likely to pay for support of services deployed on their network **even if they are created by external developers**. A developer who can understand and mitigate this risk for the Mobile operator stands to get attention from the operator.

A service could go through the following phases:
- Concept/idea – the initial vision
- Validation through market research and analysis through focus groups, research reports etc
- Prototype creation
- Prototype acceptance after unit testing
- Full development of the application
- Deployment of the service
- Promotion
- Customer care, billing etc

Thus, we see that there are other components which supplement development and all must be in place for a successful service. In addition to the factors listed above, the developer should also consider issues like
- Digital rights management
- Legal and statutory issues which may differ depending on country of deployment
- Billing mechanisms
- Copyright, partnership agreements, license agreements
- Pricing
- Competitive analysis
- Demo versions, promotional versions
- Customer relationship management
- Brand
- Unique selling proposition

A new service starts with an idea. Historically, in any field, the percentage of ideas which can be successfully translated into businesses are low. In the dot-com era, this rate was even lower. The main characteristic of dotcom ventures was the low cost of deployment and the low impact of failure. This is not the case with the Mobile data industry. Extending the example of game development we have seen before, testing for compatibility on handsets is a significant cost – and this testing has to be carried out on each supported handset for each operator.

Depending on the idea, being first to market may not necessarily translate into business success. The combination of relatively high cost (compared to the web) means that it is necessary to validate ideas. Thus, market demand coupled with customer segmentation should be the starting point rather than a specific technology.

Even if the concept is completely different, traditional methods such as market research, customer focus groups etc can add value to identify the demand for the new service. Because the cost of deploying a new service is not trivial – the Internet approach of 'put it out there hoping that customers will discover it' does not translate well in the Mobile data industry. Creating a new service and simply leaving it without a formal demand driven strategy backed by promotion, means we are depending on 'luck'. The chances of success are low. Worse, the downside is also more severe since it costs more to create a service than it does to create a traditional web site.

Hence a more formal approach is called for. In the simplest analysis, we recommend the following model to identify the target customer base. **The overall principle is – we start with the entire demographic and narrow down to smaller subsets of people some of whom we could convert to customers.** The questions which drive this model are

- What is the overall population you are addressing? This could be the population of a country, a continent, a region, a city etc
- What is the percentage of handset penetration amongst this population
- Which operators are you targeting within this population?
- Which handsets are you targeting within this population?
- What is the technology of deployment for example Java, SMS, WAP etc? (These are explained in more detail below)
- Does the application have any special technology needs such as location-based services? How many people have handsets equipped with this technology?
- What does a segmentation analysis of the subset now reveal (see below for a discussion on segmentation)?

- What are the distribution channels to market for the segments we are targeting?
- What proportion of this subset do we expect to hit and convert to customers based on our marketing budget?

Thus, your target audience subset times number of potential downloads per month should give you an idea of your monthly revenue. This could then be tied against your cost base including your development costs, porting costs etc to arrive at a more tangible picture of success/failure of the new service.

Conventional wisdom says that SMS, WAP, Java, BREW are the technologies which have achieved mass market in at least some countries. However, as we have shown above, the issue is wider than a specific technology. Two aspects of the above model need further discussion – i.e. customer segmentation and distribution channels to market.

3.5.2 Customer segmentation

Building on the discussion above, whether your service is radically different or traditional, it's necessary to understand the target audience. The process of segmentation classifies/categorises your customer base into groups so that your marketing messages and channels can be focussed towards your target audience. Segmentation and marketing are significant areas in themselves. We are looking at segmentation from an innovator (and not the Mobile operator) perspective. However, because the service is likely to be deployed via a Mobile operator, it is necessary to understand what factors are important to the Mobile operator. Closely related to segmentation, is the topic of 'channels to market' i.e. the channel to market must be chosen to fit the customer segment.

Traditionally, telecoms segmentation was simple i.e. based on factors such as age, sex etc. With the advent of more services, segmentation is becoming more complex. Some dimensions along which customers could be classified are:

a) Business vs. non business customers
b) Customers in a specific vertical market – for example financial services
c) Customers by billing mechanism – prepay vs. post pay
d) Segmentation by age of the customer
e) Segmentation by sex
f) Segmentation by type of handset technology used

g)   Segmentation by geography etc

3.5.3 Distribution channels to market

Distribution channels facilitate access to the customer and are closely tied to the specific segments we wish to target. In fact, the largest distribution channel is the Mobile operator itself. The operator has a direct relationship with the customer. The developer is, in effect, using the operator as a channel by selling to the operator's customer base. However, the operator is not the only route to the customer. Other channels could be used to deploy your application. In addition, the operator, as a channel, is also undergoing some transformation, as we shall see later in the section on 'phased opening of the gardens'.

Potential application distribution channels include:

a)   Embedded applications within the handset – for example embedded games like 'Snake' on Nokia handsets. This requires a partnership with the device manufacturer

b)   Partnership programs created by operating system vendors such as Symbian. (www.symbian.com). In general, many large vendors have partnership programs. Often, these programs also have a 'marketplace' i.e. a place where third party applications can be sold.

c)   Third party content sites on the Internet for example www.monstermob.com

d)   Distributors who could sell games or content from their site for example www.mforma.com or in turn resell the content to operators with whom they have built commercial relationships.

e)   Selling applications in high street stores – either Mobile operator owned or independent stores

f)   SIM applications – i.e. applications deployed on the SIM card

g)   Developer programs from handset manufactures such as forum Nokia http://www.forum.nokia.com and Ericsson mobility world http://www.ericsson.com/mobilityworld/

h)   Integrated developer programs such as from Qualcomm/brew as explained below http://brew.qualcomm.com/BREW/en/

i)   Bundling and packaging with other products - for example CD inlays for selling ringtones

j)   Media / promotion / advertising companies who may manage brands. Promotion campaigns for these brands could include a Mobile element. Examples of such agencies are www.flytext.com
k)   Web portals such as yahoo www.yahoo.com
l)   Other media like Newspapers, television, radio stations etc who could be potential partners due to their access to a large customer base.
m)   Specific channels depending on application types – for example offline dating agencies for Mobile dating applications
n)   Systems integrators who could have existing relationships with large corporate customers.
o)   Third party Mobile portals either based on WAP or more recently Java.

3.6 Partnership

From the above discussion, we see that there are two pitfalls that any new service must overcome – firstly the service needs to be demand driven even if it is a totally new service and secondly the right set of partners should be considered **especially partners who are channels to market.**

The term 'partner' is used loosely in many business relationships. In its purest sense, a partner is someone who shares both the rewards and the risks. The best example we can think of is the 'Wintel partnership' i.e. Wndows and Intel. In this partnership, the partners share both the upside and the downside. Both have clear demarcations – in this case Intel for hardware and Microsoft for software. The relationship is 'trusted' such that beta versions of code are shared, future products are planned in consultation and so on.

Ideally, the innovator wants to be a 'partner' but could become a 'supplier'. A supplier is not a partner because a supplier does not share risks. A close relationship with suppliers is important to the success of a business. This is even more so in businesses like Automotive manufacturing where parts are obtained as close to production with the 'just in time concept' and retail (for example Walmart(www.walmart.com), Tesco(www.tesco.com)) where inventory is acquired as late as possible from the supplier so as to carry less stock. Nevertheless, suppliers do not share risks and hence are not the same as a partner when it comes to sharing the rewards. Suppliers can be changed easily – partners less so.

Partnerships need not be with the Mobile operator directly. In the Mobile data industry, the question of 'who to partner with' is crucial and explored in the chapter 'Inside the mind of the Operator'. Developing the application is just the first step but but it's much more difficult to make it into a commercial service unless you align with the right partners. Not only should partners be (ideally) channels to market but also their interests must be in tune with yours. We will explain this in more detail in subsequent chapters.

Some questions to consider when deciding potential partners are
- What skills do we bring to the partner?
- Is the partner's sales cycle compatible with our own?
- The market is currently fragmented both in business terms (for example interconnect agreements not in place between Mobile operators) and technology terms (lack of uniform standards).In this scenario, what percentage of the market does the partner cover?
- What are the branding arrangements? Is your brand visible to the end customer or does the partner's brand dominate the whole service?
- What are the billing arrangements?
- What are the revenue share arrangements?
- How is your application positioned in terms of the overall product offering to the end customer? For example – your application may just be a means to showcase the partner's infrastructure.
- Is your content/brand creating the demand or is the demand being driven by the partner's marketing/portal?
- How unique is the content/application?
- What is the level of capital investment if any required from you and the partner?

The obvious path is to deploy the application through the Mobile operators. However, as we have seen from the distributor channels above, it is not the only path. For a small developer with just a few applications, an aggregator is also a good option. Further, there is nothing preventing a developer from selling applications directly from their web site provided they could build up sufficient volumes.

Most developers approach Mobile operators through their developer programs. Developer programs are evolving and today (end 2004) we are seeing a lot more progress in this space. From the early days, the Mobile operators were sending out the right signals about partnerships. But the reality was different. Certainly, the cards were (and still are) stacked largely against the small developer in the current model. In the early days, some Mobile operators had developer forums. In principle, they were avenues for the small developer to gain access to the Mobile operator. In practice, they were often euphemisms for selling the Mobile operator's services (for example - training and certification). Due to financial considerations, most Mobile operators had given up trying to seriously engage with the 'garage' developer – choosing instead to focus on the safer branded content or to work with a few large developers - which is a risk averse strategy.

The industry hasn't been exactly kind to us, the developers. Under pressure from adverse market conditions, major players have changed direction 180 degrees, often forgetting the poor developer along the way. Leaving aside new companies that had developer programs and went belly up (like www.omnisky.com) large, existing vendors are equally immune. Our favourite example of flip-flops (i.e. U-turns) is Motorola (www.motorola.com). Motorola started as a Symbian/Java evangelist on Mobile devices, then suddenly dropped Symbian ('flip'), adopted Microsoft platforms ('flop'). Currently, some of their phones in the marketplace such as E365 support J2ME i.e. Java while others such as MPx200 support Microsoft sending confusing signals for developers. As Ed Zander (ex president of Sun's software group) becomes the chairman and CEO of Motorola, we wait to hear another 'flip' towards J2ME.

Thus, there have been many teething problems but that is changing as the economy picks up, the industry becomes more mature and the scope for commercially successful applications increases.

So far, the only well documented partner model to date is the NTT/Japanese model. The key elements of the NTT model from a developer perspective are:
a)   A revenue share of approximately 90/10 in favour of the developer
b)   Two classes of applications are possible – the official DoCoMo applications have a billing relationship and a set

of conditions (i.e. a legal agreement with NTT). In contrast, the unofficial sites are left to their own devices but many are still profitable.

c)   NTT has significant clout and has created assembly line processes around the content creation process (la 'Toyota'). Their clout can influence even the vendors such as Sun Microsystems to influence Java standards.

While this model has been a success in Japan for a few years now, there is little evidence that it is being replicated outside of Japan. In fact, we believe that it has some drawbacks and **unlikely** to be adopted in Europe and USA where the 'wholesale model' is favoured by operators which we describe in the next section.

3.7 MVNOs and the wholesale model

Developers often gaze fondly at the greener grass across the oceans – in this case Japan and Korea at the 90/10 model of revenue sharing. Why don't we have a 90/10 share ask many?

However, an often ignored characteristic of the Japanese model is the fact that the Mobile operator is not sharing revenue from carriage at the wholesale level at favourable rates. In fact, due to the entrenched model that already exists in Japan now – we believe this is unlikely to happen in the future. In contrast, Europe has much better wholesale rates with a wider range of price points. This means, the European model is more suited for MVNOs (Mobile virtual network operators) and the rise of MVNOs is already in evidence in both Europe and USA.

A MVNO buys network capacity on a wholesale basis from a Mobile operator and provides services to a customer base using that capacity. Thus a MVNO does not run its own network but rather 'piggybacks' on the infrastructure of an existing Mobile operator. The nature of the MVNO can range from a simple 'no frills' service to managing a part of the network function such as a dedicated Mobile switching centre/roaming etc.

MVNOs broaden the overall market base since we now have more channels to market for content and applications. While the I-mode model has opened up the market in a certain way, the rise of MVNOs will create a different business environment in Europe and USA. Thus, the 90/10 arguments do not translate directly to Europe and USA because it depends on the baseline i.e. '% of what'? - combined with what functions the MVNO takes on (marketing, customer care, billing etc). This is not to say either model is better or worse – only to observe that they are different and potentially mutually exclusive i.e. given one the other is not likely to arise in a specific geography.

MVNOs offer benefits from the cost side as well as from the services side by pushing new services to their existing user base. According to AT Kearney (www.atkearney.com), MVNO margins could average between 20 to 40%. Our expectation is that net returns look slimmer at 4 to 6% but could be greater for branded portals. Thus, the wholesale/MVNO model is likely to prevail in Europe/USA as opposed to Japan where the wholesale model does not lend itself to creating a business by third parties such as MVNOs.

Hence, we see a rapid growth rate on the European MVNO scene compared to the Far East. German MVNO Mobilcom http://www.mobilcom.de/ announced a 7.6% revenue increase in Q4 – 2004 over the same quarter a year ago. In the UK, the partnership between Virgin (www.virginMobile.com) and T-Mobile (http://www.t-Mobile.com/) is already a success story and is drawing more entrants. The virgin group lends the brand along with managing the customer relationship whereas T-Mobile provides the Mobile network infrastructure. Already, virgin customers account for 4 million subscribers of the total 14.4 million T-Mobile UK subscribers.

But there are other reasons for the growth of MVNOs. Operators running a 'fixed only' Mobile network have an incentive to take the MVNO route to overcome the threat of 'Mobile only' vendors stealing a portion of their fixed line revenue. And some 'Mobile only' vendors can address new markets using the MVNO strategy for example the Miami based Tracfone http://www.tracfone.com which specialises in the Hispanic market partners with 'Mobile only' operators such as Verizon (www.verizon.com) and Cingular (http://www.cingular.com/)

Many new entrants are entering this space – especially brands such as Easygroup (www.easy.com) and Disney (www.disney.com). From a Mobile operator's point of view, the relationship with MVNOs brings mixed blessings and depends on the leverage the MVNO brings to the table – especially the degree of branding. Partners such as Virgin can manage the marketing and customer service saving the Mobile operator customer acquisition costs and ongoing customer care costs. By focusing on niche audiences (such as the example of Tracfone above), MVNOs enable Mobile operators to save on the costs of formulating multiple marketing strategies for niche markets. For the MVNO, there could be some disadvantages for example not being able to manage the quality of the network. But the overall advantages outweigh the disadvantages. Retail brands, media companies, niche market specialists like financial services companies etc all are potential candidates for MVNOs. Thus, MVNOs are strategically important to the innovator because they broaden up the overall marketplace.

Many innovative cooperative models are emerging where MVNOs and Mobile operators can collaborate. One example is emerging UK based MVNO 'Subzone' which is a student based operator. Subzone aims to reduce churn at a critical point in a young person's life for example - when they leave college. Thus, the student may start at college with Subzone but as their circumstances change – they may adopt different MVNO brands but while still staying with the same Mobile operator. (Because the operator will progressively market different MVNO brands at critical life changing junctures – such as leaving college). We could call this model Mobile DIVA – 'Direct and Interactive Value Acquirer'.

This collaboration works because both MVNOs and Mobile operators are working on the same business assumptions that are - users spend a certain amount of money on their phone services (ARPU) and that they change services periodically (churn). A collaboration that increases ARPU and reduces churn is beneficial to both.

3.8 Partnership, portals and revenue shares

The issues of partners, revenue shares, Mobile operators and walled gardens all come together with Mobile portals. The concept of portals comes from the fixed line Internet with sites such as www.yahoo.com, www.aol.com etc. A portal is viewed as the 'starting point'. Due to this role, any application that is at the top of the portal gets maximum exposure to the customer. Positioning on the portal is controlled by the portal owner (generally Mobile operator). Revenues are collected by the portal owner and shared with the services on the portal. The positioning and revenue share as applied to Mobile portals is driving factor behind walled gardens as viewed by the industry as a whole today. As we shall see later, this is a one-dimensional view of the walled garden i.e. it assumes that the portal is the only 'channel' which it is not.

Even the Mobile portals themselves have many elements and differ considerably from their fixed line counterparts. Many people actually do not have first hand experience of Mobile portals. Also, in many cases, operators charge to just to browse the portal, which can get expensive. The UMTS forum (www.umts-forum.org) categorises Mobile portals into five Divisions i.e. **Mobile Intranet/Extranet, Customised Infotainment, Multimedia Messaging Services, Mobile Internet and Location-Based Services**

Some of this categorization is already outdated. Already, we find that some of these classifications are out dated. Multimedia messaging was a 'hot' technology a few years back. Today, its role is far more uncertain – let alone it's role as a portal. In contrast, we find the advent of the new Java based portals (as opposed to WAP portals) more interesting since they allow more players to interact directly with the customer. (Java and WAP technologies are explained below). Java can be downloaded as an application from any web site. Thus, if I am a regular reader of a content source – say the Financial Times –then I could download the 'Java application' from the financial times web site. Because the content on this application can be updated through an Internet connection, I can get a similar experience as that of a portal.

There are many factors that can influence a portal – personalization being the most important. This is because users are not likely to spend a lot of time browsing the portal from a small Mobile device especially when they may have to pay for browsing! Other considerations for Mobile portals according to the UMTS forum include Markup Language (for example XHTML Basic), User Capabilities, Mobile Terminals including (Browsers, Operating System, Chip Technology, Battery Technology and Display Technology), Security (USIM, Privacy), Billing (Flexibility Billing, Multiple payment Methods), Quality of Service, Interoperability and Content Formats such as MP3.

Revenue shares are negotiable and vary globally. In Europe and the USA, they depend on some of the factors we have highlighted before for example who is bringing more to the table (existing brands would command a higher revenue share). The three reference points are: the Japanese model (with 90% to the developer), credit card model (the Mobile operator keeps 2.5% to 4% just like a credit card company) and finally a 50/50 split. Ultimately, it all depends on building a viable business. Operators may want 50/50 or even more but if a developer cannot sustain a business at that level then everyone loses! On the other end of the scale, the existing brands want a lot more for their brand but do not realise that at these levels, the operator will find it unsustainable.

This comes back to our argument for the need for innovation, assuming that as a garage developer you do not have an existing brand and/or extensive funds – you must differentiate somehow.

3.9 Our thoughts

We summarise our thinking below - some aspects, which we have already discussed, but others, which we will articulate more below
   a) Based on the discussions, we believe that a pragmatic two-stage approach is needed. Firstly – to identify what services innovators can focus on now. Secondly to experiment with services/technologies that have potential to create genuine competitive advantage for the grassroots innovator in the near future.
   b) We believe that the Mobile data industry is historically consistent in its evolution i.e. it's similar to any other industry. Its future (comprising of utility based services) should not be confused with the present (mainly content based applications).
   c) We discuss some technologies and models in the forthcoming sections, which developers should consider (for example Parlay X) and the OpenWaspa model. But we also discuss principles of simple services that could make money right now.
   d) We believe that without an 'innovative component', new services will not succeed in this industry (especially when they are not backed by strong brands, extensive funding etc). Thus, innovation is necessary – both for the new entrant but also for the industry as a whole to thrive.
   e) We believe that opportunities exist 'on the fringe' by combining one or more elements to create a new service.
   f) The phased opening of the walls has the potential to create many opportunities as we discuss below

g) Partnerships are crucial with each party bringing something useful to the partnership. A partnership is different from a supplier relationship.
h) The service must be demand driven

4 Chapter Four: Technical Landscape

So far, we have explored the Mobile data industry from a strategic perspective. Our key question has always been – 'How can a thousand flowers (entrepreneurs) bloom on the Mobile Internet? I.e. as a developer, how can you build commercially successful applications in the Mobile data industry? Following on from our previous discussions, we look at the current technical landscape. This discussion will form the basis of our outline on the 'plug and play' ecosystem in the next section.

Even if you do not have a technical background, we urge you to read this section. It discusses components like web services, which are important to understanding the ultimate OpenGardens vision.

On the other hand, if you have a technical background, you might find this section 'not detailed enough' and that's intentional!

The Mobile data industry follows the same standards as the Internet. There are good books that discuss specific technologies in detail and we do not wish to replicate them. Thus, if you are already familiar with technologies like J2EE, XML etc. – you are half way there. For a more complete discussion of technology, we recommend the book 'Next Generation Wireless Applications' By Paul Golding (Wiley).

Our goal in discussing technology here is to build a foundation to the 'plug and play' (OpenGardens) ecosystem we have alluded to before. Thus, from a technical perspective, we only discuss technologies that are necessary in the understanding of the OpenGardens model.

4.1 Accessing the Mobile Internet

In this section, we discuss methods by which Mobile clients can access information. These include client side technologies like WAP (Wireless Application protocol), Java, BREW, XHTML, Microsoft windows technologies etc. Before we start discussing client access technologies, we first look at Mobile applications models and acquire a basic understanding of the Mobile Internet.

4.1.1 Mobile applications models

As an applications developer creating applications external to the Telecoms network (i.e. third party applications when viewed from the point of view of the Mobile operator), there are number of possible models –

Browsing applications:
Conceptually the same as browsing the web but taking into account limitations unique to mobility (for example - small device sizes)

Messaging applications:
That can be classed into two groups –

> **SMS** (Short messaging service) and **MMS** (Multimedia messaging service), which originated within the Telecoms infrastructure and

> **IM** (Instant Messaging), which has its roots in the Internet world.

For now, we can view SMS and MMS as sending text or multimedia messages between two entities in a 'store and forward' mode. In other words, SMS and MMS are non real time (although they often appear to be 'real time'). IM on the other hand is a presence based, synchronous messaging system operating in near real time i.e. a connection is first formed between the two communicators and messages are sent within that session. Thus, the two communicating entities have to be physically present at the time of communication (synchronous).

Downloading applications:
Downloading applications are applications that are first downloaded and installed on the device. The application then runs locally on the device. Currently, most Java based games seen today are downloaded applications.

Embedded applications:
Applications that are capable of running natively on a Mobile device operating system (such as Symbian applications – see below)

SIM card based applications:

Applications that are capable of running on SIM (Subscriber Identity Module) cards

4.1.2 The Mobile Internet

Misleading as the phrase is, the 'Mobile Internet' is still a good place for us to start our discussions. To start with, we can view the Mobile Internet as extending the conventional (landline) Internet over the air interface (also called RF network for Radio frequency network). The underlying protocols used in the Mobile Internet are the same as the Internet i.e. **IP (Internet Protocol)** and **HTTP (Hypertext transfer protocol)** or its WAP variants, which for all intents and purposes are now pretty much the same as HTTP. From a technical standpoint, the Internet can be viewed as linking IP aware machines with each other.

From a user standpoint, the Web is a series of web pages accessed via a browser. A browser can display any information formatted in **HTML (Hypertext Markup Language).** A collection of HTML pages is the familiar **World Wide Web.** For most part, the Internet operates in 'client server' mode – the client making the request via the browser and the server fetching the content and serving it in HTML. On the server side, the Internet introduces a new element called **'Web server'.** In response to a request from the browser, a web server extracts information from data sources, formats the results in HTML and returns the results to the browser. HTTP is used to communicate between the browser and the web server. The basic Internet interaction model is as depicted below.

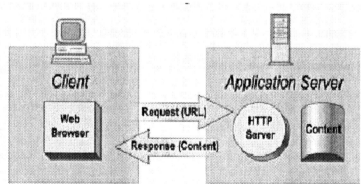

Source: Wireless Application Protocol Architecture Specification WAP-210-WAPArch-20010712 www.openMobilealliance.com

The combination of HTTP and HTML power the Internet. There has been considerable momentum behind the Internet and Internet protocols are now mature and well known – both from a technical and developer perspective. It would be useful to extend the same Internet/World Wide Web protocols on to the Mobile Internet. While it would be ideal to use HTTP/HTML over the Mobile Internet, it is not optimal to do so since they do not take into account mobility related factors such as small device size and unreliable connections over the air interface, etc. The **WAP forum** (now incorporated under the open Mobile alliance www.openMobilealliance.org) has studied these issues and sought to address them in context of the RF network. Where possible, the WAP forum has sought to use Internet standards like HTTP/HTML as a foundation in creating new standards/protocols for the Mobile Internet.

4.1.3 WAP (Wireless Access Protocol)

Overview

In the previous section, we mentioned the WAP forum. The WAP forum was responsible for governing the much-maligned **WAP (Wireless Access Protocol).** Much has been said about WAP – most of it not flattering. However, if we see the work of the WAP forum in two parts i.e. WAP as a user interface and WAP as a transport mechanism – then the picture changes significantly. WAP as a user interface (i.e. as a browsing mechanism) never really took off mainly due to hyped up marketing leading to misguided customer expectations. However, WAP as a transport mechanism is an outstanding success and is now an integral part of the Mobile Internet.

By 'transport mechanism' we mean the use of WAP in solving the problems outlined above (the deficiencies of HTTP/HTML over the air interface) while at the same time being able to communicate with standard HTTP web servers (because we want to conform to the Internet development protocols).

The WAP proxy (or WAP gateway) is the mechanism to solve the problems of optimal communication over the RF network. As its name suggests, the WAP proxy maintains a standard HTTP connection with the web server on behalf of the Mobile device. Thus, we end up using HTTP but at the same time 'not using it' over the RF network as shown below

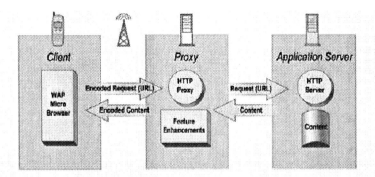

Source: Wireless Application Protocol Architecture Specification WAP-210-
WAPArch-20010712 www.openMobilealliance.com

The protocol used by the WAP proxy to communicate between the Mobile device and the WAP proxy depends on which version of WAP we are referring to. WAP itself has undergone two major modifications – WAP 1 and WAP 2. To communicate over the RF network, WAP 1 uses a protocol called Wireless transport protocol **(WTP) and WSP** (Wireless session protocol). WAP 2 uses Wireless profiled TCP (Transmission control protocol) (**W-TCP)** and wireless profiled HTTP **(W-HTTP).** W-TCP and W-HTTP are much more closely aligned to the standard TCP and HTTP but with modifications that are acceptable extensions to the HTTP/TCP specifications. Since we now have a mixture of devices some using WAP 1 and others using WAP 2, newer Mobile devices support both protocols WAP 1 and WAP 2.

WAP push
In addition to the browsing model, WAP offers an additional communication paradigm called 'WAP push'. WAP Push is traditionally used for notifications and alerts. WAP push can be deployed over a number of bearers including SMS. In fact, a WAP push notification along with a WAP download link is commonly sent over the SMS bearer to deploy content. Once the user clicks on the WAP download link, an IP session is established and the content is downloaded. WAP push is also the delivery mechanism for MMS.

4.1.4 XHTML-basic

We mentioned before that WAP is an excellent transport mechanism but not widely used as a browsing mechanism. The technology favoured for browsing applications is XHTML-basic. XHTML-basic is a version of XHTML aimed at small information devices such as Mobile phones. XHTML is an encoded version of HTML4. The rationale behind XHTML is to achieve conformance with XML (extensible mark-up language). XML is described in detail later – but for the purposes of this discussion, we can view XHTML-basic as a modularised version of XHTML. In other words, XHTML is compliant with HTML and XHTML-basic is a modularised version of XHTML suited for small devices.

4.1.5 Java and J2ME

Although WAP started off as the preferred method of developing client side applications, today Java from Sun Microsystems (www.sun.com) along with BREW (discussed below) are the preferred method for client side application development. Java differs from XHTML because the application is downloaded and runs on the client (as opposed to XHTML, which acts as a browser). Java is also a favored mechanism for developing Mobile games because games need a richer development environment, which Java provides.

Java comes in three flavours – J2EE (Java 2 Enterprise edition) for the enterprise/distributed environment, J2SE (Java 2 standard edition) for the desktop and J2ME (Java 2 Micro edition) for small devices. J2ME can be viewed as a subset of J2SE, which in turn can be seen as a subset of J2EE. We are concerned with J2ME (http://java.sun.com/j2me/) in this section. J2ME is "modularised" to cater for a range of devices. Modularisation in J2ME is achieved via configurations and profiles.
The complete Java platform is depicted as below.

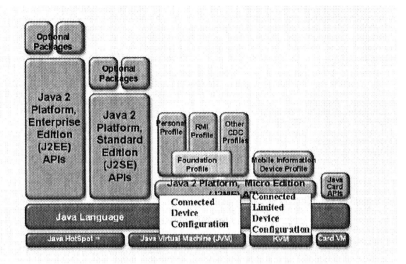

Source: www.sun.com

A **configuration** is aimed at a horizontal group of devices for example devices with similar memory size and processing power. It defines the minimum Java libraries and VM (virtual machine) capabilities that a developer can expect to find on all devices implementing the configuration. A **profile** is a collection of Java API's that mould a configuration to provide industry-sector agreed-upon capabilities. A profile is concerned with a specific archetypal device family suited to a particular market for that industry sector. Several combinations of configurations and profiles are possible.

For low end PDAs and Mobile phones, CLDC (Connected limited device configuration) applies. For high end PDAs, the CDC (Connected device configuration) is applicable. The MIDP (Mobile Information Device Profile) profile is applicable to Mobile phones.

MIDP (http://java.sun.com/products/midp/) is a profile within J2ME designed for Mobile phones and entry-level PDAs. MIDP 2.0 (released in Nov 2002) is the current specification of MIDP and is backward compatible with its predecessor MIDP 1.0.

MIDP 2.0 is a significant leap forward for Mobile devices. It brings a number of new features such as secure networking, the Push model and OTA provisioning.

Secure networking: MIDP 2.0 mandates that HTTPS (secure HTTP) be supported in all compliant handsets. Prior to this, secure networking was not a standard feature of MIDP 1.0

Push model: MIDP 2.0 includes a server push model whereby MIDlets can be registered to be activated when a device receives information from a server.

OTA provisioning (Over the Air provisioning) is now required as part of the MIDP specification. OTA provisioning defines how MIDlet suites are discovered, installed, updated and removed on Mobile devices.

Sun's mantra of 'write once run anywhere' is not quite true with Mobile devices. The larger Mobile operators have used Java but tried to change it to their own needs for example i-appli used with i-mode. Although I-appli is a Java environment, the JCP process (i.e. Java community process) does not govern it. In contrast, the other two Mobile operators in Japan (KDDI and J-phone) are using Java, as we know it. Similarly, device manufacturers can also influence MIDP by extending the profiles i.e. providing additional features (APIs) specific to their own device. This is very much a part of the Java design but defeats the purpose of write once run anywhere.

4.1.6 BREW (Binary runtime environment for wireless)

BREW http://brew.qualcomm.com/BREW/en/) from Qualcomm (www.qualcomm.com) is a development environment for client side applications on CDMA devices.
BREW competes with Java but both coexist in their respective environments. BREW is the only option for Mobile operators who have adopted CDMA technology. In practice, this means North America and Korea. For other Mobile operators, Java is the preferred option. In spite of being a 'closed' environment, BREW is actually favorable to developers since it addresses the entire value chain in a 'one shop' mode. Developing in BREW involves developing in 'C' or 'C++'. Qualcomm manages the entire relationship with different Mobile operators including revenue shares, billing etc.

This factor alone has the potential to make BREW a more viable proposition for developers.

4.1.7 Microsoft technologies

Throughout this book, we have not referred to Microsoft in depth. This is due to our focus on the Telecoms network/consumer/web services model. In a Mobile environment, the Operating system (Microsoft's forte) is not as critical as it is on the desktop. Microsoft's strategy is based on its strengths in the corporate world (PDA's) rather than the consumer side (phones). In this role, we expect it to enjoy considerable success but limited leverage in a wider world of non-PC centric devices.

4.2   The Telecoms network

Having approached Mobile applications from the foundation of the Internet, we now discuss the core components of the network itself.

4.2.1 The Telecoms RF networks

A Mobile radio network can be classified into two broad categories – The Mobile operator managed **RF network** and the **localised networks** created around hotspots/access points. The latter includes technologies such as Bluetooth and wireless LANs. We first look at the Mobile operator managed RF network. We will revisit RF networks again in subsequent sections since the RF network is essentially being 'Opened up' as part of the OpenGardens philosophy.

The Mobile operator managed RF network is a 'cellular' service in the sense that the actual network can be viewed as a honeycomb of 'cells'. A cell is a basic geographic service area of a wireless Telecommunications system. Cells are created by a large number of low power transmitters. This results in a honeycomb like structure of cells. An idealized representation of a cellular network is as shown below (note neighboring cells do not use the same frequency)

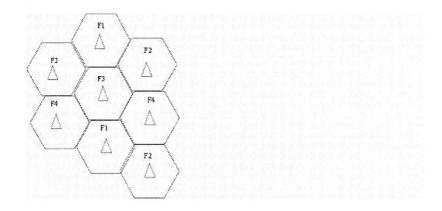

Technologies like SMS are application level technologies. They are in turn based on underlying cellular data transmission technologies (i.e. the technology governing the cellular/RF network). There are two main techniques for cellular data transmission – **TDMA (Time division multiple access) and CDMA (Code Division Multiple Access).** The objective of both techniques is to support multiple, simultaneous data channels.

TDMA achieves this objective by dividing the radio frequency into time slots. TDMA is used by the GSM cellular system (described below). **CDMA** (Code division multiple access) comes from a Military/Defense background and is currently used by major cellular carriers in the United States. Qualcomm largely patents it. CDMA uses a more complex mechanism to support simultaneous data channels, which is out of the scope of this book.

Note: The objective in all cellular transmission techniques is to distinguish the signal from individual transmitters. For simplicity, we have chosen to discuss only TDMA and CDMA since they are used in cellular systems worldwide. There are other methods for cellular data transmission such as FDMA (Frequency division multiple access) and PDMA (Polarization division multiple access). For more information on cellular data transmission techniques, please refer http://en.wikipedia.org/wiki/Cellular_network

These technologies have evolved into different **cellular systems** across different geographic locations. Cellular systems can be viewed as 'generations' of systems as they evolved over time (2G, 2.5G, 3G etc.). The main differentiation across the generations is support for greater bandwidth. Obviously, as you go towards 3G and beyond – the bandwidth increases and the applications supported also become richer. First generation (1G) were analogue systems. From 2G (second generation) onwards, the cellular systems have been digital.

2G systems

GSM (Global System for Mobile) is the most popular 2G system. GSM originated in Europe and is the dominant Mobile system across the world. In some form, it is present in all continents including North America. GSM (based on TDMA) is a digital system with a relatively long history (the study group was founded in 1982) and is governed by the GSM Association (http://www.gsmworld.com/index.shtml). The GSM Association provides functional and interface specifications for functional entities in the system but not the actual implementation. Besides GSM, other examples of 2G systems are cdmaOne (mainly in the USA), PDC (Personal data cellular) in Japan.

2G technologies are typically capable of supporting up to 14.4Kbps data. 2G systems are characterized by being **'circuit switched'** (i.e. a circuit is first established between the sender and the receiver before sending the information and is maintained for the duration of the session). The next evolutionary step (2.5G described below) is characterized by being **'packet switched'** (the data is broken into packets and no connection is maintained for the duration of the communication). Note that both circuit switched and packed switched networks are digital.

2.5G systems

2.5G networks are an intermediate step undertaken by most Mobile operators in their evolution from 2G to 3G. The main functional leap between 2G and 2.5G networks is the adoption of packet switched technologies (in 2.5G networks) as opposed to circuit switched technologies (in 2G networks). 2.5G networks are capable of theoretically supporting bandwidth up to 144kbps but typically support 64kbps. GPRS (General Packet Radio Service) in Europe and CDMA2000 1X in North America are examples of 2.5G networks. Applications such as sending still images are possible over 2.5G networks.

3G systems

Most people know about 3G in terms of the high prices paid by Mobile operators for 3G licenses. However, from a Mobile operator perspective, there is a clear business case for investing in 3G because existing 2G networks are congested 2.5G solutions are a half way house and will not cope with the increasing demand (i.e., both the number of consumers and the richer application types such as Video).

From an application perspective, 3G technologies are differentiated from 2.5G technologies by a greater bandwidth (theoretically 2Mbps but typically 384kbs). Possible 3G applications include video streaming.

From an end user point of view, the move from 2.5G networks to 3G networks is more evolutionary than revolutionary except in the case of devices. 3G devices are significantly more complex because of the need to support complex data types like video, provide more storage, and support multiple modes. **UMTS (Universal Mobile Telecommunications System)** in Europe and **CDMA2000** in the North America are examples of 3G systems. Note that 3G systems are all based on CDMA technologies.

4.2.2 The Bluetooth personal network

Bluetooth is a wireless technology specification, which enables devices such as Mobile phones, computers and PDAs (personal digital assistants) to interconnect with each other using a short-range wireless connection. It is governed by the Bluetooth SIG (special interest group) at www.bluetooth.org

The operative word is 'short range'. A typical Bluetooth device has a range of about 10 meters. The wireless connection is established using a low-power radio link. Every Bluetooth device has an in-built Microchip that seeks other Bluetooth devices in its vicinity. When another device is found, the devices begin to communicate with each other and can exchange information. Thus, a Bluetooth enabled device can be thought of as 'having a halo' seeking to communicate with any device that enters the range of that halo.

The significance of Bluetooth is – it is 'free' in the sense that it is an extension of the IP network in an unlicensed band via the Bluetooth access point. Although Bluetooth has not lived up to much of its initial hype, the technology is significant since most phone manufacturers have committed to Bluetooth enabled phones.

From an application development perspective, Bluetooth can appear in many forms for example:

- As a **technology for redemption** of coupons i.e. a marketing coupon could be sent over Bluetooth and redeemed at an access point in the store.
- As a **payment mechanism** – a Bluetooth Wallet can be a secure payment mechanism.
- As a **'location based' service** since location is known within range of a Bluetooth access point.
- As a **mechanism for forming ad-hoc contacts** via bluejacking (http://www.bluejackq.com/)
- **Bluetooth communities** such as (http://www.bedd.com/)

Bluetooth is often compared to WiFi technologies (see below). The two technologies operate in the same frequency range (2.4G). Functionally, they achieve different things. Bluetooth, in its minimal form, is a cable replacement system operating in a point-to-point mode. WiFi, in its minimal form, is wireless networking (i.e. Ethernet, or point-to-multipoint). Both technologies coexist.

Bluetooth unfortunately has been hyped up a lot by the media and often suffers the consequences. It does what it is supposed to well (i.e. at a minimum, act as a replacement for cables to synchronise a range of devices) and in that role will always be useful.

4.2.3 The WiFi personal network

Wireless LANs is a term, which refers to a set of products that are based on the IEEE 802.11 specifications. The most popular and widely used Wireless LAN standard at the moment is 802.11b, which operates in the 2.4GHz spectrum along with cordless phones, microwave ovens and Bluetooth.

WiFi enabled computers or PDA (personal digital assistants) can connect to the Internet when in the proximity of an access point popularly called a 'hotspot'. WiFi differs from the cellular system in the use of the unlicensed frequency (where as the cellular systems always operates within a licensed spectrum). Access points have a limited range for transmission - around 100 meters (328 feet) indoor and 300 meters (984 feet) outdoors.

The Wi-Fi Alliance (http://wi-fi.org/OpenSection/index.asp) is the body responsible for promoting WiFi and its association with various wireless technology standards. Recently, a group of 14 Mobile operators, handset and equipment manufacturers joined together to release a new open specification designed to facilitate WiFi - cellular handoffs on dual mode handsets http://www.umatechnology.org/. Currently, this relationship pertains to telephony and not data.

Note:
A cellular handoff is the seamless connection between WiFi and wireless. In practice, it means your Mobile phone will first seek a WiFi connection. If one isn't found, it will go to the air interface. If you later roam into an area with WiFi it will then drop back to WiFi. Thus a 'seamless handoff'. This does not exist today but it is forthcoming.
For WiFi, also see the wireless ecademy
 http://wireless.ecademy.com/index.php chaired by Tony Fish.

4.3 Location

4.3.1 Overview

Location i.e. 'M' for mobility, as we have seen before, is the 'Missing 'M'' and also the chief differentiation for mobility based applications. In user terms, a location based application would deliver any content filtered by location – for example, a dynamic map that mirrors a user's location or a 'find my nearest' application where every fast food outlet in a one kilometer radius can be listed. Although location is one of the main drivers/differentiates for wireless - Europe and USA have different motivations for Location Based Services (LBS).

In the USA, the motivations for LBS are statutory. The Federal Communications Commission (FCC) issued a mandate to all US wireless carriers that required them to implement Enhanced 911 (E-911) services for all wireless users. Essentially, E-911 services help to position a caller to the emergency services. In its complete form, the accuracy of the location would be down to 100m. Once such accuracy is available, it can lead to more applications apart from the positioning of emergency services.

In Europe, there is no statutory requirement on the carriers. Hence, the market drivers are commercial only i.e. value added consumer services.

By definition, any Mobile service is 'location based' – however currently the accuracy of the location is only to the 'cell level' and the size of each cell is not uniform. It is the greater accuracy (for example, 100m) within a cell that can enable new applications not currently possible.

According to Mobile commerce (http://www.Mobilecommerce.co.uk/), the top 10 searches for Mobile information are

Top 10 Searches

Source: http://www.Mobilecommerce.co.uk/corporate/didyouknow.htm
reproduced with permission

4.3.2 Features of a Location based service

An end-to-end view of a location-based service includes:
- The application that the consumer will interact with from their handset
- A means to determine the location of the subscriber's handset (positioning technologies)
- A means of managing location requests from the Mobile operator's side

- A way of mapping that location on to content, such as a map

- A method of delivering content back to the consumer, filtered by their location

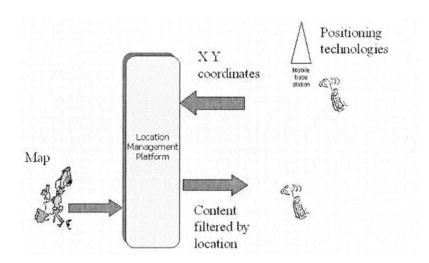

As a first step, the Mobile operator identifies the location of the subscriber's handset. The Mobile operator is the only entity that knows the user's location (unless the Mobile operator releases that location to third parties that take location feeds from one or more Mobile operators). The Mobile operator runs software, which provides the platform to implement location-based services.

This location is then mapped onto a content feed that the platform may take from a number of content providers (such as a mapping company). The resultant information is sent back to the end user through the application. In the above transaction, clearly there is value for the end user since they get location enabled, timely information.

Possible location scenarios include:
- User initiates the location by sending a message/query first – active location based services

- The Mobile operator (who knows the user's location) initiates the transaction passive location based service

- A third party requests the user's position when the user has previously agreed (opt-in) to receive messages. It then sends a message to the user based on a profile that it holds.

In simplest terms, the location is the 'x and y' coordinates of the subscriber known to certain accuracy. However, two aspects make all the difference

a) The accuracy of the location (i.e. position determination techniques) and

b) The active/passive modes of location determination

Merely determining the 'x-y' coordinates is not enough.
To provide a full value added service, three elements are involved:

- **Position Determination** which return the 'x-y' coordinates - i.e. techniques to understand where the user is located

- **Location Management platforms** software that interfaces with the position determining techniques providing independence from a specific positioning method and

- **Geographic Information enablers** – mechanisms to translate the location coordinates into information useful to end users such as street addresses.

We can look at these aspects in more detail.

4.3.2.1 Position determination techniques

Determining location comes down to knowing coordinates (i.e. positioning). To know the position of a Mobile device within a network, either the same signal must be read by three different fixed stations (call Triangulation) or the position of a Mobile device must be known within a coverage area (say, the coverage area of a cell).

There are broadly two types of positioning techniques: **Terminal based and Network based.** Besides, we have hybrid combinations of the above two methods.

Terminal based solutions:
Terminal-based solutions depend on modification/processing through the terminal. These include:

- **GPS**: Global Positioning System uses a network of satellites to locate a user's position. Although used for a while in vehicle navigation systems, this technology is only now being used with Mobile devices.

- **A-GPS:** Also uses the satellites as in GPS but complements the satellite readings by using fixed GPS receivers that are placed at regular intervals.

- **E-OTD** – Unlike GPS, Enhanced Observed Time Difference (E-OTD) mainly uses software in the terminal but still requires investment from the Mobile operator. The E-OTD procedure uses the data received from surrounding base stations to measure the time difference it takes for the data to reach the terminal. That time difference is used to calculate where the user is located relative to the base stations.

Network based solutions:

As the name suggests, pure network-based solutions do not need modifications to the terminal.

- **CGI-TA** - Cell Global Identity (CGI) uses the identity that each cell (coverage area of a base station) to locate the user. This is complemented with the Timing Advance (TA) information. The accuracy of this method depends on the cell size and the smaller the cell size, the better the accuracy. Accuracy ranges from 10m (a micro cell in a building) to 500m (in a large outdoors macro cell). This is a good and a cheap solution for proximity type services (find my nearest) where high accuracy is not required. The caveat is - accuracy depends on the size of the cell.

- **TOA** – Similar to E-OTD, Uplink Time of Arrival (TOA) works by measuring the uplink data (the data that is sent by the terminal). The key difference between E-OTD and TOA is that TOA supports legacy terminals. However, a Mobile operator needs to upgrade all the base stations making the overall cost of deployment very high.

Other methods

- **Bluetooth** (www.bluetooth.org) can act as a location-based technology within a small confined space -around 10m to 100m around a Bluetooth transmitter.

- **Cell broadcast** is a Mobile technology that allows messages to be broadcast to all Mobile devices within a designated geographical area (normally a single cell). More information can be obtained at http://www.cellbroadcastforum.org/

Note that application developers will not interact directly with positioning technologies. From the perspective of application design, any positioning technique essentially returns the coordinates through software API, which are received by the location management software and mapped on to the content. The choice of technology is based on factors like accuracy, regulatory issues (E-911), frequency of update, latency and device availability. In the development of location-based services, a useful resource is the http://www.locationforum.org (now part of the Open Mobile alliance www.openMobilealliance.org)

4.3.2.2 Location management platforms

This is typically a Carrier grade solution from companies such as Webraska (www.webraska.com). We do not discuss location management platforms here in detail since they are typically installed by the Mobile operator and developers can treat them as a black box as long as they can interface to them.

4.3.2.3 Geographic Information enablers

Location based services are often related to mapping since they often involve overlaying location information on maps. Using mapping functions, a developer can request a specific section of a map to a given scale. Maps can be held in either vector or a raster formats.

- **The vector format** stores features as points, lines or polygons. A point is a single x- y coordinate, a line is a series of connected points and a polygon is a series of connected lines closed to form an area feature. In user terms, a point may be the location of a 'Point of Interest' (POI), a line could represent a road and a polygon could represent a sales area.
- **The raster format** maps are created by actually scanning the paper-based maps. Most web-based maps utilize this format and the end user easily understands them. However, bandwidth limitations imply that they are not the best format to be used with wireless.

Since information is already stored in the vector format, maps created using the vector information are more efficient. Once an address is known, the process of **geocoding** converts the address to an x-y coordinate. Often the user wants to query something in relation to the current address (**find my nearest**). However, while an 'as-the-crow-flies' spatial search is relatively easy to compute, it may not be practical from the user's perspective. For example, the topography on the straight route may include a mountain or a stream. Thus, the directions have to be converted to a route by a process called **routing**, which provides directions from one geocoded point to another. Routing is not static i.e. it takes into account a range of factors, such as traffic updates. Traffic updates can change, so within a journey, the route may be recalculated.

4.3.3 Location based services – the issues

While everyone agrees that location is critical to the success of mobility, there are still many challenges to be overcome. The two most important are privacy and location technique.

Privacy and legal

Legislation/privacy is a concern with respect to location bases services. For example employers can track employees. This means employees will have to consent to being tracked.

Similarly, children's' charities feel not enough is done because in a 'locate a friend' service - the person doing the tracking can remain anonymous. Consent is deemed to be via a text message that the 'trackee' accepts from the tracker. This scheme is open to abuse especially if the tracker's identity can remain anonymous (i.e. merely having the trackee accept the tracker's invitation to track may not be safe enough). In other countries, there are some initial examples of Location based services. In Japan, buddy services are popular where alerts are sent when your friends are in the same zone. In Finland, you can send location based text messages, which will be picked up when other users enter the same area.

Location based Mobile dating is another application being considered but some Mobile operators have lumped the 'dating' applications along with other 'adult' applications and seek to impose the same restrictions on them.

http://www.followus.co.uk/ and http://www.mapaMobile.com/ are examples of tracking users (with their consent)

In addition, the information collected by Mobile operators is also subjected to legal restrictions. Mobile operators cannot provide location information to other parties. However, the demarcation becomes unclear with regards to government. How much information can the government acquire and for what purpose?

In current technology (typically some form of cell broadcast i.e. locating a device within a cell), you can locate a Mobile device within 500m in a built up area. However, that will change soon with GPS where you can narrow down to 10-50 meters. The privacy stakes are high in this scenario.

In the UK, the Mobile operator '3' is providing location based services via GPS but due to privacy concerns and the need to authenticate users, three has chosen to avoid third party location services.

Active tracking vs. passive tracking
Leaving aside concerns for privacy and other legal issues, an ideal application would be one where the Mobile operator, in real time, passively tracks the user. The feed is then provided to a third party that reacts, assuming the user has given their consent. Thus, the third party may action a communication accordingly – for instance, when the user walks past a store, into a mall or enters a certain postcode or cells (assuming that they have the user's permission to contact them in this way).

Besides the passive tracking of location, it is possible to track location in an 'active mode' where the user's location can also be provided by the user themselves, if they wish to disclose it to a third party. For instance, the user could see a poster advertising say, a car and SMS their current (or home) post code to the advertiser to find their nearest dealer. Clearly, this puts the onus on the user, but can be highly effective if the user wants the information, or is provided with a sufficient incentive to request it.

For technical reasons, the full-scale passive tracking of location (tracking everyone at all times in the network) is a challenge. Full scale passive tracking of users involves tracking all users, at all times relative to fixed or moving points of interest when the user themselves could be moving. This is very challenging!

Passive services are more feasible when the 'trackee' gives permission to another person/entity to track them. These include services like Buddy finder, child-minder, etc. There is already some instance of such services as we referred to above.

4.4 Mobile commerce

4.4.1 Introduction to m-commerce

According to Durlacher (http://www.durlacher.com/), *"Mobile Commerce refers to any transaction with monetary value that is conducted via a Mobile telecommunications network."* Mobile commerce (m-commerce) makes sense, intuitively. A large number of purchases are made on impulse and m-commerce is ideally suited for impulse purchases. Unfortunately, as with many other aspects of wireless, m-commerce too was widely hyped.

Mobile Commerce is driven by:

- **Ubiquity** – anytime, anywhere purchase. For example - Cinema, theatre or airline tickets

- **Time critical purchases** (promotions and special offers)

- **Purchase of entertainment type services** (for example, games)

- **The prospect of 'Micropayments'** i.e. getting customers to pay small amounts between 5p to £5. At these levels, credit cards are not viable on account of their commission charges (normally a minimum charge of around 3%).

- **The prospect of Mobile 'Coupons'.** Magazine/Newspaper coupons have been used for a number of years but with the advent of m-commerce, coupons have become much more interesting. Coupons also fit in logically as part of Mobile marketing. Typically, the coupon is sent as part of the marketing campaign on the Mobile phone and is redeemed for a discount at the point of sale.

A Mobile Commerce model ideally favours products that the user does not need to see, touch, feel or try on before purchasing, as they are in a standardized format. Examples might be books, CD's, DVD's, videos or tickets. Mobile Commerce comprises two components - **Mobile Shopping** and **Mobile Payments**. Mobile shopping is concerned with the dialogue with the customer and helping the customer to place the order. Mobile payments, on the other hand, deals with issues of actually handling the payment such as security and credit card fraud.

The market for Mobile commerce transactions has a range of players each with their own strengths and weaknesses:

- **Mobile operators** are experts in managing networks and have a large subscriber base but lack the capabilities to develop compelling applications with speed and scalability. Mobile operators also do not necessarily have the marketing skills and brand management skills needed to connect the right customers with the right products. Further, they do not have the capacity to handle payments (in the way that banks can). Mobile operators are looking at Micropayments where they have leverage (typically, payments between 5p and £5).

- **The banks** can handle payments but often not micropayments since they lack the technical infrastructure to handle micropayments unlike Mobile operators. They are interested in this market since they see the Mobile operators trying to encroach on their traditional business.

- **Credit card companies** – are interested in the market for the same reasons as the banks. Credit card companies are also interested in using their existing network effectively.

- **Handset manufactures** – have promoted the 'dual-chip' handset approach where the bank supplies a second SIM (subscriber identity module) in addition to the Mobile operator SIM that is then used for m-commerce. Currently, there is some focus on m-wallet technology.

- **Cooperative scenarios** – These are collaboration efforts between the above parties for example - the Mobipay platform in Spain is available to all four Mobile networks. Other scenarios are possible - for example, between one Mobile operator and one bank (e.g. Postbank and Telfort in the Netherlands) or one Mobile operator and several banks (e.g. the Mobile operator TIM and Italian banks in Italy).

- **Pure play vendors** – Such as Paybox who are solely in the business of providing a Mobile payment platform.

Back in the heady days of 2001, a study conducted by MORI, Britain's largest independent market research agency along with Nokia (www.nokia.com), found that the number of people interested in using m-Commerce is more than eight-fold in some markets compared to the number of people actually using eCommerce today (2001).

From the study:
http://press.nokia.com/PR/200105/822176_5.html
The study also showed that nearly 90% of people interested in using m-commerce services will also be willing to pay extra for the convenience of making purchases with m-commerce.

The study revealed that Mobile-phone users view m-commerce as complementary to alternative remote-commerce channels such as the Internet.
Specifically, the study found that initial adoption of m-commerce is likely to be on a similar scale as today's usage of eCommerce, which is already somewhat mature. Also, 24% to 54% of respondents across the markets stated they would be willing to carry out a transaction of more than USD 25 using a Mobile device.

They tend to favour "local" transactions, where m-commerce provides a unique application for electronic transactions. Their choice of payment method depends on the size of the transaction and billing arrangements, but most are willing to pay extra for m-commerce services. This suggests that consumers see real value and benefit in using their Mobile phones as tools for shopping.

The study demonstrated that convenience and control will play a pivot role in acceptance of m-commerce. Study participants saw m-commerce as a way of avoiding carrying cash or waiting in queues, as well as a way to grain greater control over expenditures and enjoy unlimited purchasing possibilities.

In 2004, how many m-commerce transactions have you conducted? What went wrong? Hype apart, m-commerce faces many hurdles – legal, privacy, willingness to pay extra for m-commerce, standardisation and so on. The present state of M-commerce is not stable or mature. The sector displays the classic symptoms of an emerging industry (fragmentation, immature legislation, lack of standards, lack of critical mass etc.). The biggest problem is 'creation of value'. M-commerce is widely perceived to be 'nice to have'.

4.4.2 Implementation of m-commerce systems

There are three ways to implement an m-commerce system

1. Through a **Mobile operator led initiative** like Vodafone m-pay

2. **Third party platform** led initiatives like the now defunct 'Paybox' or

3. **Premium SMS**

The first two are attempts to create a standard and have largely been unsuccessful. Premium SMS on the other hand has enjoyed considerable success.

Consider the case of Paybox, which approached m-commerce from the banking perspective reflecting their background of banking (Deutsche Bank backed the company). Paybox connected the Mobile phone to the bank so that it was possible to pay for services and products through the phone and also transfer money to other people using the phone (A person to person model). To pay for a service using Paybox, users had to first register with Paybox. When they wished to settle a bill, the retailer enters the consumer's Mobile telephone number into the existing payment system - which was integrated with Paybox - instead of their credit or debit card number. Then Paybox calls the user's Mobile phone to request a four-digit PIN to authorise the transaction.

Once authorised, Paybox informed the retailer, then debited the user's current bank account as with a normal direct debit payment. The user automatically received a text message as a receipt of the purchase, and a further email confirmation is an option. The retailer was also sent a message as confirmation of the payment authorisation. With the benefit of hindsight, there were a number of problems but the most significant was - retailers had to integrate into the Paybox system. It needed a significant body of retailers to make the system useful. Further, how many such systems should retailers integrate into? At what cost – and to what benefit?

We will touch on this problem later in the book since it illustrates the OpenGardens method.

Predictably, other initiatives came along. For example - in March 2004, Orange, the Mobile operator, ran a Mobile coupon trial offering a two-for-one ticket deal at over 90 percent of the movie theatres across the UK every Wednesday night. The problem in both these cases is the same. The retailer has to now integrate into a number of new systems each claiming to be the next 'standard'. The battle as we see below is at the 'vendor/retailer' first before we even get to the consumer.

Contrast this with the only success on the m-commerce front – Premium SMS. The success of premium SMS was due to its simplicity and its ubiquity. Premium SMS involves a special type of SMS message that is chargeable. There are two ways to charge for a message either on the outgoing message or the incoming message (Note – here 'outgoing' and 'incoming' are from the perspective of the Mobile operator)

Premium SMS MT (Mobile terminated)
The charging is done on the outgoing message. Premium SMS MT is preferably used when a return-SM is sent to the Consumer. Typical usage scenarios include applications where content is consumed or delivered within the SMS. Also, Premium SMS MT is highly suitable for subscriptions (one registration, many premium deliveries).
Example: Ringtone-purchase, subscriptions and alerting services.

Premium SMS MO (Mobile originated)
The charging is done on the incoming message. Premium SMS MO is preferably used when a receipt or return-SM is not sent to the Consumer. Typical usage scenarios are applications where the content is not consumed in SMS. Example: WEB-content, Voting/games without Mobile return-receipt.

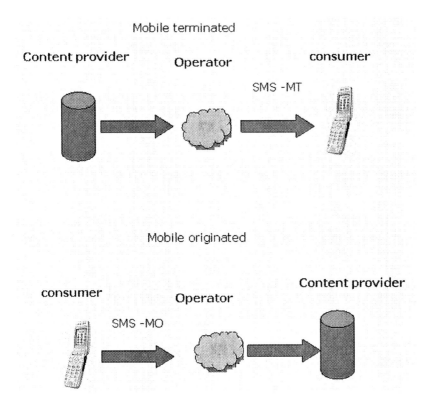

The success of Premium SMS in many markets is encouraging but it is used mainly to purchase items such as Ringtones. M-commerce as a whole remains yet to be fully exploited. For example – while the promise of impulse purchasing adds value to the consumer, the lack of standardisation from a consumer/retailer perspective will hamper development.

4.4.3 E-money regulations

Finally, before we leave this topic, in UK/Europe the FSA (Financial Services Authority) and EC (European Commission) are drafting new regulation, which affects the uptake of m-commerce (called e-money regulations). The basic premise is – If consumers are using a prepay phone (a majority of the market), then they must only buy items that are consumed on the phone.

This means, they can buy ringtones etc. but not a bottle of soft drinks from a vending machine. Apparently, it covers all aspects of payment – premium SMS, IVR, Bluetooth etc. The only way out of this regulation for Mobile operator - is for the Mobile operator to buy a banking license. At the moment, none of the Mobile operators have a banking license. Taking on a banking license would mean that Mobile operators would have to conform to a number of regulations (quarterly filing with the FSA, setting aside a percentage of the revenues for bad debt etc.).

In reality, many countries (and many in the EU itself) are taking a much more liberal attitude towards m-commerce. In Finland, you can conduct a range of transactions on your phone such as paying for trains and in the UK itself, the London congestion charges can be paid by SMS and so on.

Mobile operators are taking different strategies. Vodafone (www.vodafone.com) is presumably looking to be a factoring company – taking on debt from third parties. Some Mobile operators are also working on an initiative called Simpay (www.simpay.com) that acts as a clearinghouse for Mobile payments across Mobile operators. Thus, the picture is uncertain at the moment. The FSA consultation paper is at
http://www.fsa.gov.uk/pubs/cp/172/index.html

4.5 Messaging – SMS, MMS and IM

We briefly discussed messaging before in contrast to browsing. We discuss messaging in greater detail below.

SMS was the first messaging technology to emerge. SMS started as simple person-to-person text messaging over GSM. In this format, SMS was popular because it was simple and ubiquitous. Nokia, created a proprietary extension to SMS called **Smart Messaging** which effectively kick started the Mobile content market such as ringtones and picture messages. Many of the pioneering features of smart messaging are now commonplace (for example (Over The Air OTA service configuration). MMS was seen as the ultimate evolution of SMS with the capacity to support sound, pictures, video etc. Depending on the underlying bearer, MMS could be viewed as emerging in two phases

The first phase being based on 2.5G technologies, like GPRS, as a bearer. In this case, the MMS message looks like a short PowerPoint presentation on your Mobile phone (i.e. a series of simple "slides" featuring colour graphics, pictures, animation and sound). The second phase being MMS on 3G. Once 3G is deployed, more sophisticated features like video could be introduced.

Some companies adopted an evolutionary path to MMS by adopting a standard called **EMS** (Enhanced Messaging Service) which did not require them to upgrade their network infrastructure. Unlike EMS, for MMS to be deployed, the network Mobile operators have to upgrade their infrastructure and devices supporting MMS must be available.

4.5.1 SMS

The Short Message Service (SMS) is a means to send and receive messages to and from Mobile phones. SMS started within the GSM network standard and can incorporate words, numbers and limited binary formats.

The success of SMS worldwide defies conventional marketing wisdom. Consider the following: the people who spend most on SMS messages (teenagers) would be considered 'cash poor' i.e. not a high value customer base. SMS was never actively marketed. Instead, in some circumstances (such as prepay with SMS), it was actually discouraged initially. In spite of the huge uptake, the cost per message has not dropped leading to high revenue generation for players in the value chain. The success of SMS leads credence to the belief that content is NOT king ('contact' is!)

The working of SMS is based on a few important principles:
- **Store and Forward** SMS messages are transferred between devices via a Short Message Service Centre (SMSC). An SMSC is software residing in the Mobile operator's network and manages processes like queuing the messages and billing. SMS is thus always a store-and-forward system since the message is never sent directly from the sender to the receiver but always passes through an SMSC. If the message cannot be immediately delivered, it is stored until the subscriber becomes available or eventually it's deleted from the system.

- **Limited to 160 characters** A single GSM SMS is limited to 160 characters of user data for 7-bit default alphabet, 70 characters for non-Latin alphabets and 140 bytes for binary data.

- **Use of signaling path** GSM uses the signaling path rather than the main radio channel. This implies it works in parallel with voice calls.

- **Compression and Concatenation** From its inception, the 160-character limit has been restrictive. There are two ways to overcome this limit: SMS concatenation and SMS compression - both of which have been standardized by the GSM SMS standards.

   Concatenation - Concatenation essentially strings together SMS messages to enable the creation of one large SMS message. In theory, up to 255 short messages can be concatenated. Currently, Mobile operators do not offer a different tariff for concatenated messages i.e. a concatenated message costs the same as a non-concatenated message. SMS applications are also not geared towards handling large volumes of information. Therefore, in practice, concatenation beyond three or four messages is not practical.

   Compression - SMS compression is yet another attempt to overcome the 160-character limit. Compression and Decompression may take place between the devices that send/receive messages or between a device and the SMSC (note that the device may not be a phone – it could be a computer capable of sending/receiving messages).

- **Binary SMS** - An SMS message can be rendered in two forms – either as a plain text message or as an 8 bit binary format. The User Data Header Indicator flag defined on the message indicates the mode of data in the message. Binary messaging must be supported by the SMSC – but most SMSC's currently support binary messaging.

4.5.2 MMS concepts

Originally, many saw MMS as an evolution of SMS. MMS has some similarities but also differences with SMS. Like SMS, MMS also works on the 'store and forward' concept. The equivalent store and forward engine for MMS is the MMSC (Multimedia messaging centre).

However, although related to SMS, MMS is very different from SMS in terms of implementation. In fact, MMS is similar to email in the sense that MMS can be viewed as a presentation layer over the basic email protocol. MMS message encapsulation is based on **MIME (Multipurpose Internet Mail Extensions)** which are the same as for email. MMS messages can be sent from a Mobile device to a user who will receive it as email. Also, email clients can send an MMS message to a user. This is helpful for the uptake of MMS since it creates an audience for MMS messages even when the initial uptake of phones is limited.

Key MMS concepts include:

- **Support for SMIL** SMIL (Synchronised Multimedia Integration Language - pronounced 'smile') is used as a language to create MMS messages. In SMIL, presentation information is coded into a presentation file. The intention is to present the Multimedia content in a specific order at a pre-determined interval. SMIL is a specification of the World Wide Web Consortium (W3C) and has widespread industry following. SMIL is similar to HTML in its syntax and constructs and is essentially a way of choreographing rich, interactive multimedia content for real-time delivery over the web and also over low bandwidth connections. SMIL delivers multimedia presentations consisting of elements such as music, voice, images, text, video, and graphics all synchronised across a common timeline (i.e. not delivered as attachments). An example of an SMIL multimedia layout file consists of a news video, emphasising specific news stories with text headlines, and displaying, for example, a stock ticker at the bottom of the screen. SMIL can be viewed as a 'PowerPoint-style' presentation on the Mobile device. Using a simple media editor, users can incorporate audio and video along with still images, animation and text to assemble full multimedia presentations.

- **No attachments** Unlike an email, an MMS message has no attachments. The whole message is one entity.

- **In-built support for multimedia presentation:** MMS was built from the ground up to support a Multimedia presentation. HTML does not provide timing information for Multimedia presentation and separate media components have to be downloaded separately.

- **Delivery** Like SMS, MMS is based on the 'store-and-forward' principle.
  E-mail delivery is based on a Store and user Retrieve (S&R) service. There is a mailbox in the e-mail server and only e-mail headers are delivered to the recipient. The actual message is fetched only when the user chooses a specific header.

- **Storage** MMS messages can be stored in three places: At the **MMSC** before the message can be delivered to the recipient terminal OR in **the memory of the terminal**, after the terminal has received the MMS message OR in a **separate permanent space,** where the recipient has moved it for storage.

4.5.3 MMS applications

Apart from direct person-to-person messaging, there are two areas where MMS applications could be developed: On the **client side** i.e. for the device or as **external applications** to the MMSC.

An external application is an application that interfaces to the MMSC. An MMSC treats messages received from such applications as similar to messages sent by an end user. This is where you, the developer, will be creating your application. In this situation, the developer is referred to as a **VASP** (Value Added Service Provider). VASP connectivity would enable third parties to create applications like:
- Location based advertisements as Multimedia Messages
- News updates
- Content conversion etc.

The standard that defines connectivity for between the VASP and the MMSC is called the MM7 interface / standard from the 3GPP. Typically, the MMS application request could be triggered by sending an SMS or via a web site (such as for a weather report). The MMSC receives the request and communicates that request to the VAS application. The VAS application packages the content and the SMIL and sends the content to the subscriber via the MMSC. From the perspective of the MMSC, MMS applications can be originating, terminating or processing.

Originating applications sends an MMS message to the MMSC. The MMSC treats the message in the same way, as it would have if it had received it from a terminal. Examples of such applications are - A web based MMS creation service that can compose an MMS message.

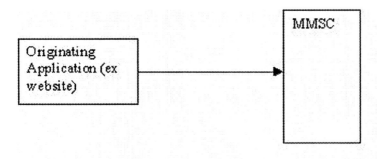

Terminating applications – are the final destination of the MMS message. An example is a photo store application. An MMS user has taken a photograph and sends it to the MMSC, which redirects the MMS to the photo store application.

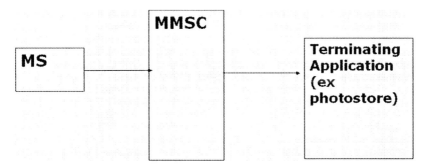

Processing applications receive an MMS message, perform some processing and send the message to the MMSC for further processing as outlined below:

Step 1 Sender sends message to MMSC

Step 2 MMSC sends message to processing application

Step 3 Processing application sends message back to MMSC after processing

Step 4 MMSC sends processed message to receiver

For example a company could subsidise an MMS message by adding a logo to the message. In this case, the MS sends a message to the MMSC and the MMSC redirects the message to the application, which adds the content and the logo and then sends it back to the MMSC to be forwarded to the receiver. Other examples of processing applications are: address verification and content conversion.

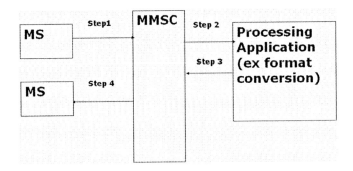

4.5.4 MMS – The real world scenario

From a developer's perspective, MMS has been a disappointment because the Mobile operators have yet to open up their MM7 interface to public access. For once, unlike WAP, devices supporting MMS were available but third parties i.e. who are not Mobile operators, could not send messages to these devices. This meant that except for a few chosen partners, a majority of the messages were person to person – and these too have been sporadic in the case of MMS.

Because of this, there have been attempts to bypass the Mobile operators in the deployment of MMS. Technically, MMS can be viewed as simply WAP push followed by WAP download. A WAP push notification sent over SMS is used to indicate to the user about a waiting message and WAP download is used to download the message. As long as the device supports MMS itself along with WAP push and WAP download, it is theoretically possible to send an MMS message to the device via WAP. In theory, this is true. In reality, it is a 'smoke and mirrors' solution. For starters, it is possible to block WAP push altogether and Mobile operators such as Orange (www.orange.com) did just that for WAP push requests which originated from outside their network. And then there is billing. The beauty of MMS was its promise of billing for content in the same message as the content delivery message. This is not possible with the WAP solution. Billing has to be done separately, probably via premium SMS – adding to costs. Thus, this approach never quite took off and we are still waiting for MMS open interface for third parties.

In addition to the issues around MM7, MMS interoperability (ensuring that MMS messages can be seamlessly rendered across a variety of devices) and MMS interconnectivity (ensuring that MMS messages can be sent to anyone across any Mobile operator) are still not ironed out. To make MMS interoperability easier, the Open Mobile alliance (see below) has published an MMS conformance document. However, the players haven't strictly followed the guidelines and this has led to incompatibility issues.

Functionally it is possible to achieve almost all of MMS's benefits using other means (such as a Java based application). While picture messaging is an example of MMS, pictures taken using a camera phone need not be sent via MMS (for example - they can be copied to the computer or sent via Bluetooth for example). Thus, a true mass market for MMS is yet to emerge. And this may never appear. MMS could well go the way of WAP – and yet be an excellent transport mechanism.

4.5.5 Instant messaging - IM

Strictly, Instant Messaging does not come under mobility since its origins are on the Web. IM is similar to 'chat' but on a person-to-person basis (as opposed to one to many type of conversation).

Instant messaging is based on the concept of 'presence' i.e. the device recognizes when members from your 'buddy list' are online. There are no standards for IM with Yahoo, AOL and Microsoft each promoting their own system. http://www.jabber.org is the open source protocol for instant messaging. Devices such as T-Mobile's 'sidekick' support AOL instant messaging. So far, instant messaging on Mobile phones does not appear to have made a big impact on users.

4.6 Miscellaneous topics

This section covers various miscellaneous topics, which are necessary to understand the industry.

4.6.1 Standards and industry bodies

Developers have a phrase called 'RTFM' – which politely put means – 'Read the flying manual'. As a Telecoms / Mobile developer, in addition to manuals, you must also be familiar with standards which are vast in themselves. In this section, we list a set of standards bodies that you should be familiar with for development of Telecoms applications.

Because we are interacting at the cusp of the Internet, Telecoms and the Mobile Internet - there are three sets of standards you must be familiar with.
Internet standards: Which govern the Internet world
Telecoms standards: For the air interface and
The open Mobile alliance: for Mobile services.

The Internet engineering task force (IETF) www.ietf.org manages standards concerning issues such as packet routing, transport control and others relating to the core Internet itself.

Standardisation related to the air interface is managed by bodies such as the 'Third generation partnership program' (3gpp) www.3gpp.org. The 3GPP is the place to go for all 3G, GSM and GPRS standards.

Open Mobile alliance http://www.openMobilealliance.org/ (OMA) is the first port of call for all standards related to Mobile services. OMA was formed by an amalgamation of a number of industry bodies such as the WAP forum, location interoperability forum etc. This is a positive achievement and should be commended since it makes development easier for all of us.

As per the OMA web site, *the mission of the Open Mobile Alliance is to facilitate global user adoption of Mobile data services by specifying market driven Mobile service enablers that ensure service interoperability across devices, geographies, service providers, operators, and networks, while allowing businesses to compete through innovation and differentiation.*

Other industry bodies that you may need to be familiar with include
- The GSM association - http://www.gsmworld.com/index.shtml
- The Mobile data association - http://www.mda-Mobiledata.org/
- Parlay – www.parlay.org
- UMTS forum - http://www.umts-forum.org
- WiFi Alliance http://wi-fi.org/OpenSection/index.asp
- Bluetooth SIG – www.bluetooth.org
- Symbian – www.symbian.com
- BREW - http://BREW.qualcomm.com/BREW/en/
- The Mobile marketing association - http://www.mmaglobal.com/
- http://www.icstis.org/ Independent Committee for the Supervision of Standards of Telephone Information Services

Java is governed by the Java community process, which releases JSRs (specification requests). The URL for JCP is http://www.jcp.org/en/home/index

4.6.2 Serverside implementation of Java (J2EE)

As we mentioned before, J2EE (Java 2 Enterprise edition) is the enterprise/distributed implementation of Java. J2EE is widely deployed as a service delivery platform within the Mobile operator infrastructure. Prominent J2EE application server vendors include BEA (www.bea.com), IBM (www.ibm.com), Apache/Jakarta (www.apache.org)

The J2EE architecture is similar to that used in conventional enterprise/Internet environments and hence we will not describe it in detail. Instead we will look at two aspects used in the Mobile data industry - **J2EE / MVC architecture** used for browsing applications and the **J2EE/vending machine** architecture used for downloading applications.

4.6.2.1 The J2EE/MVC architecture

The J2EE/MVC (Model, View, Controller) pattern is commonly used when producing Mobile browser related applications. The MVC pattern separates the server side application into three parts – the business logic, the presentation layer and the controller to dispatch requests and control flow.

Typically, Business logic is encapsulated into **EJB (Enterprise Java beans)** and a combination of **servlets and JSPs (Java server pages)** implement the presentation logic.

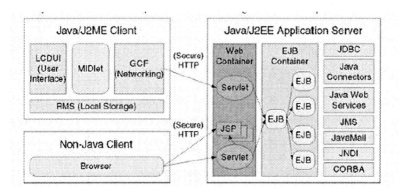

Source: Sun Microsystems
http://java.sun.com/blueprints/earlyaccess/wireless/designing/designing.pdf

4.6.2.2 The Java vending machine architecture

The Java vending machine architecture (also called a provisioning server) is described in JSR 124 (http://www.jcp.org/en/jsr/detail?id=124). This architecture is widely used when implementing content platforms.

Like conventional vending machines, the Java vending machine provides a mechanism for users to pick and choose from a range of content pieces – packaged and ready for delivery for consumption. Although deemed to be client technology neutral, the JSR vending machine architecture is optimised for MIDP i.e. Java on the client.

The overall objective is to download the application (which could be a game) in the form of a jar file. For merely downloading applications, a simple web server would suffice. However, the Java vending machine architecture is optimised to supporting a range of devices and clients with different features (such as a range of screen sizes). The vending machine can even filter out content from the menus that is not suitable for the user depending on their device capabilities.

The specification also defines a file called the Provisioning Archive (PAR) file to use for deploying provisioning applications.

Thus, the developer is concerned with creating the 'par' and the 'jar' files.

The provisioning itself comprises of three tasks:
a)   Stocking content: i.e. managing the repository
b)   Discovery: The process of finding out what content is available from the provisioning server and
c)   Delivery: delivering the content to the client

A generic provisioning server/vending machine architecture is depicted as below:

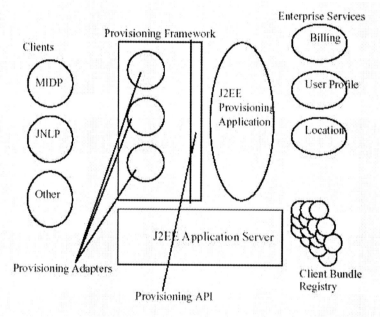

Source: Sun Microsystems: J2EETM Client Provisioning Specification Version 1.0

4.6.3 Operating systems for Mobile devices

Mobile devices have unique requirements compared to desktop operating systems such as limited battery life, unstable connections. On the desktop, Microsoft is the dominant operating system. Microsoft – through the 'Microsoft Windows Mobile OS' is still to make a large impact in the consumer Mobile space except in the PDA (personal digital assistant) segment. Symbian (www.symbian.com) is the dominant operating system in the consumer Mobile space. Besides Microsoft and Symbian, other players include BREW and Palm (www.palm.com) OS.

The software stacks you are likely to encounter are:
- BREW with C++ variants for development
- 'Microsoft Windows Mobile OS' with C++ variants for development
- Symbian with C++ or increasingly Java.
- Palm with C++

Symbian is pitching itself to be a pure 'Mobile' operating system – and it is in competition with Microsoft. Symbian is supported / licensed by Ericsson, Fujitsu, Kenwood, Matsushita (Panasonic), Motorola, Nokia, Psion, Sanyo, Siemens, Sony, Sony Ericsson and others.

A relatively new entrant is Linux on the Mobile device, which has the advantage of no license fee for the operating system.

From an openness perspective, it's easy to confuse openness of operating system with openness of applications. This simplistic view is incorrect and merely brings the desktop mentality to Mobile. Merely using Linux on the Mobile device will not lead to open applications. OpenGardens does not advocate open source operating systems but rather advocates open platforms. Mobile operators create walled gardens – not operating systems. There are increasing controls being implemented even at the Operating system level – for example Symbian applications need to be signed.

(https://www.symbiansigned.com/app/page/faq).

Unsigned applications are deemed to be unsafe. Some Mobile operators will only promote signed applications but at the moment, there are no restrictions on installing unsigned applications – which is not to say there could be none in the future.

4.6.4 Memory cards

Memory cards provide expandable onboard memory for a range of devices such as MP3 players, digital cameras and now increasing Mobile devices. The storage of content especially music on the memory card has significant potential and is already exploited by a number of Mobile vendors.

There is no defined standard in this class of products. The closest to a format not controlled by a specific vendor is MMC (MultiMediaCard) which has support from a range of manufacturers and is governed by the MultiMediaCard Association (MMCA) (www.mmca.org).

Memory cards include the following formats:
CompactFlash Card
A CompactFlash card (http://www.compactflash.org/) is a small, removable mass storage device with PCMCIA-ATA functionality. Supported by a variety of vendors from PC/portable backgrounds (such as Acer) to imaging (Kodak, canon) etc.

MMC Card
MMC cards are governed by MultiMediaCard Association (MMCA) (www.mmca.org) and are an open standard.
A number of Mobile device vendors have announced support for the MMC technology (such as Ericsson, Mobile Platforms, Hitachi, Matsushita Electric Industrial, Nokia, Sanyo Electric Siemens). Nokia has made some significant announcements for products that support MMC (Nokia 6600, 3650, Nokia N-Gage). Others such as Keitaide music in Japan, have adopted features of MMC cards. From a DRM (Digital rights management) perspective, Secure MultiMediaCards also developed by MMCA, are more interesting. These cards contain special tamper-resistant modules and incorporate the same kind of security technology used in SmartCard bank and credit cards.

SD card
Governed by www.sdcard.org. Unlike the MMC card, this is not an open standard i.e. royalties are payable. It is a secure card since it has DRM technology built in.

Memory Stick
Developed and popularized by Sony, the memory stick (http://www.memorystick.com) is used in a range of cameras, camcorders, PDAs etc. They offer DRM protection through a technology called MagicGate. A shrunk down version of the memory stick called memory stick duo is used in Sony Ericsson phones such as P800 and P900.

SmartMedia
SmartMedia cards are made of a single flash chip. They offer a low cost, portable flash solution for many digital devices. And are used in cameras, PDAs etc.

4.6.5 P2P: Peer to Peer

Most interactions on the Internet operate in the client server mode i.e. the client (through a browser) makes a request to a server that returns the requested information. In contract, the P2P (peer to peer) mode of interaction does not have a central server to store the information. In that sense, it is the opposite of client server. Because P2P does not involve central servers, the model scales faster – all things being equal.

With respect to Mobile technologies, P2P often (but not necessarily) involves Bluetooth (see section on Bluetooth for more details). In a P2P model, all nodes (computers/Mobile devices etc.) are equal and function as either client or server. Because of its decentralized nature, the peers provide most of the content on the network. Napster (www.napster.com) is the best example of a P2P network on the web.

On the Mobile front, the combination of peer to peer, Bluetooth and FOAF (friend of a friend) http://www.foaf-project.org/ are leading to new social applications. These include - lovegety (Japan), jabbwocky (http://www.urban-atmospheres.net/), smallplanet
 (http://www.smallplanet.net/), dodgeball (http://dodgeball.com/) and others. All these applications enable 'strangers' to interact with each other at a personal area network for social interactions (for example dating).

Because of Napster, P2P has always been linked to illegal file sharing. Mobile operators and content owners are concerned about repeating the Napster scenario on the Mobile Internet. Reactions to P2P vary. Korean Mobile operators have been liberal with the P2P phenomenon. SK Telecom in Korea has demonstrated a P2P client to swap pictures, music and video irrespective of whether the content is copyrighted or not. Researchers from Nokia (www.nokia.com) have come up with scheme to share content via P2P networks as part of the PerPhone project
 (http://www.nokia.com/nokia/0,,5169,00.html)

4.7 XML, Web services and OpenGardens

We now come to XML, web services and ultimately OpenGardens in the next section.

4.7.1 XML

Bill Gates called XML (Extensible Mark-up Language) the "universal canvas for the Internet Age" and believes that it will fundamentally change the way we receive and search for information.

HTML is not oriented towards exchange of information either within an enterprise - or more importantly – across enterprises. In older systems, EDI (Electronic Data Interchange) was used to address the problem of transfer of information. With the advent of the Internet, that role is taken over by XML. XML documents enable structured data to be moved across the Web. XML does not need a licence and is platform independent. It is both human and machine-readable.

The basic concept behind XML is the idea of identifying content in a document by surrounding it by tags. XML provides a language for describing content and in addition it separates the tagging of content from its presentation or style. The rules for tags are set up in the Document Type Definition (DTD). XML is thus a user-driven, open standard for exchanging data both over corporate networks and between different enterprises over the Internet. You are likely to encounter XML with an application that sends information across the barriers of the enterprise - for example, content feeds.

4.7.2 Web services

Based on XML, Web services are the current buzzwords across the development community.

Web services address a common problem – that of distributed computing i.e. communications between systems /software residing on different machines over a network. The reason for the current popularity of web services (over previous methods of distributed computing) is the prevalence of XML. This means, in principle, systems across the world developed in different platforms, could communicate with each other provided that they publish a standardised web services interface to the outside world.

This synchronisation or orchestration of diverse systems across a network is suited for applications that need to dynamically 'call' other applications owned by another business entity. For example, within the travel industry, a travel agent could access the reservation system of a car rental service directly.

> **The web services model is key to understanding the OpenGardens concept because the 'API enabling a Telecoms network' is achieved through web services.**

For a web service to be used, three things have to happen:

1. There must be a universal place where the service will be listed. This directory is known as **UDDI – Universal description, Discovery and Integration registry**

2. The service capabilities (parameters etc.) must be known to the calling entity. These are described in **WSDL (Web services Description Language)** and

3. A common language must exist to format and encode the message itself. This is achieved using **SOAP (Simple Object Access Protocol)**

The classic web services interaction model is as shown below

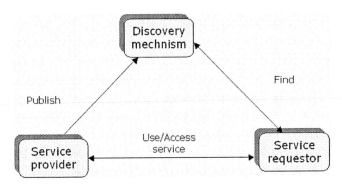

While other standards also exist within web services, these three (UDDI, WSDL and SOAP) form the basis of web services.

When viewed as a software stack, web services comprises of:
Transport layer:
The lowest layer involved with transporting messages between applications. Examples are - HTTP, SMTP, FTP etc.

Messaging layer:
This layer transports the XML messages using encoding formats such as XML-RPC and SOAP.

Service Description layer:
Describes the public interface. Implemented in WSDL.

Service Discovery:
Connects the service seekers with the service creators.
Implemented via UDDI.

Thus, a web service has four characteristics:

- **Based on XML**
- **Invoked over the Internet**
- **A public interface through which they can be invoked**
- **A search mechanism through which they can be found.**

Due to it's pervasive nature, web services are expected to be a part of most Mobile developments especially in situations where one application needs data from another, remote application.

You can find examples of web services at
http://www.xmethods.com/ and
http://www.alphaworks.ibm.com/webservices
The IBM site shows you a simple example created by combining stock quotes, traffic reports etc. into one simple 'portal like' site. Note that these feeds are from other business entities and are dynamic.

We will return to our discussion on web services later. Web services form a key component of the OpenGardens philosophy through initiatives such as Parlay X.

5 Chapter Five: Devices

In this section, we use the word 'device' to refer to all kinds of Mobile devices including phones, PDAs etc. The specific features of devices are changing rapidly and can be compared easily through any mainstream buyer's guide. (In the UK, we used www.what-cellphone.com). Hence, rather than discussing features, we discuss trends.

There are a number of ways to classify devices –for example dedicated devices which perform one task only (a basic Mobile phone used to make phone calls), an integrated device which combines more than one function (for example smartphones like Sony Ericsson P900 which can be used to play MP3 music), modular devices (like Nokia n-Gage which have optional add ons such as memory cards for a specific game). Devices can also be classified according to technology for example CPU, memory, power consumption, radio frequency, operating system and so on.

> **Our list below gives a set of features, which we believe are important to the UK/North American market. Although this analysis/methodology is UK centric, you can use the same classification to conduct your own analysis specific to your geography. You can even derive stats and figures based on this analysis i.e. by starting with market share you can then extrapolate percentages of phones supporting certain features.**
>
> **However, it's important to look at the big picture – for example delivery of Video on both 2.5G and 3G technologies – rather than just 3G. Use this book as a guideline to isolate your own critical factors.**

Here are the trends as we see them as at Sep 2004

5.1 Top manufacturers and their market share

According to www.idc.com , worldwide Mobile phone shipments in 2Q04 increased 2.5% over the first quarter and rose 36.7% year-over-year to 163.7 million units.

Rank	Vendor	2Q04 Unit Shipments	2Q04 Market Share
1	Nokia	45,400,000	27.7%
2	Motorola	24,100,000	14.7%
3	Samsung	22,700,000	13.9%
4	Siemens	10,400,000	6.4%
4	Sony Ericsson	10,400,000	6.4%
5	LG Electronics	9,940,000	6.1%
	Other	40,800,000	24.9%
	Total	163,700,000	100.0%

Notes: Vendor shipments are branded shipments and exclude OEM sales for all vendors.

In spite of still maintaining the market leader position, Nokia had lost considerable market share to the Korean vendors Samsung and LG electronics

5.2 Converged -v- non converged devices

Will users use one device for many functions or will they use different devices each specialized in a specific function? The Nokia n-gaga www.n-gage.com was a seminal device since it was the first large-scale attempt to combine a gaming device plus a phone. Other variants are possible – An iPod (http://www.apple.com/iPod/) with a phone and so on.

An alternate model, which we believe is more feasible – is a Personal Mobile Gateway (PMG) whose best proponent is IXI Mobile (www.iximobile.com). In this model, each device is optimized to perform it's own function but different devices are all capable of connecting together through the personal Mobile gateway which is a small device either built into the cellular phone or carried separately.

5.3 The importance of form factor

In spite of all the technological additions, the basic device form factor is often the driving factor. The market leader Nokia learnt this the hard way when they did not introduce a 'clamshell' device until much later in comparison to manufacturers like Samsung

5.4 The trend towards richer devices

The trend towards richer devices is no surprise (for example devices supporting colour). Even when devices support colour, there is a trend for making the colours more vivid (for example Nokia 3510i supports 4096 colours and functions as a low end gaming device and the Nokia 3660 supports 65K colours). However, there still exist a large number of 'legacy' devices in the marketplace.

5.5 Support for WAP

Support for WAP is almost mandatory in most devices but only the newer devices (such as Nokia 6600) support WAP 2.0 whereas relatively new but popular devices (such as the Nokia 7210) still support WAP 1.2

5.6 PDA -v- smartphone

In the early days, PDAs like the Palm, Pocket PC etc. were clearly distinguishable from phones (such as the then popular Ericsson T68i and the Nokia 7110). Today, with devices like the Sony Ericsson P900, that distinction is increasingly blurred. In this sense, devices are indeed converging.

5.7 The blackberry

The blackberry from Research in motion (www.rim.com) has built a reputation on its unsurpassed ability to manage email. It has displaced top end corporate devices like Nokia 9210i to be a preferred choice among corporate executives.

5.8 Phones supporting WiFi

Devices such as HP iPAQ h6340 Pocket PC support WiFi in addition to GSM/GPRS and Bluetooth. Support for WiFi is a welcome change http://www.umatechnology.org/ for standards to support GPRS/WiFi handoff

5.9 Who 'owns' the customer?

Customers do not like to be owned but companies like to think they own customers. Although the Mobile operator manages the connection, the device manufacturer is the first brand name the customer sees (i.e. Sony Ericsson, Samsung etc.). Mobile operators however are getting into the game of manufacturing/ influencing devices. For example – Vodafone in the UK has successfully promoted Sharp GX 10, GX 20 and GX 30 exclusively. In Korea, the Mobile operator SK Telecom is actually looking to buy handset manufacturers against regulatory disapproval!

5.10 Support for EMS

Surprisingly, as at Sep 2004, there are still phones that support EMS such as Alcatel Onetouch 332. EMS, for all practical purposes, is a dead technology – amazingly someone still buys these phones!

5.11 Location

Hardly any phones support location, which should act as a warning to the proponents of location based services. Only the newer 3G phones such as Motorola A835 support location via A-GPS. Note that others may support location via software means and mechanisms such as cell broadcast. However, these have limitations as we have seen above.

5.12 Support for Java

Is widespread but not universal. There are still many devices that do not support Java but most high-end devices do. Still, relatively few phones support MIDP 2.0 such as Sony Ericsson P900 and Nokia 6600. For a good list of MIDP 2.0 phones, refer http://www.benhui.net/modules.php?name=Midp2Phones

5.13 Support for memory cards

Support for memory cards is fast growing as a standard as content becomes richer. Phones like Motorola MPx200 have SD/MMC card slots. Others like Motorola A835 have a 64Mb internal store but no slots for memory cards. Sony Ericsson P800/P900 supports the memory stick. Memory cards have varying DRM protection schemes as explained above.

5.14 Support for Symbian

There are really two choices for operating system – Symbian and Microsoft. Support for Symbian is widespread amongst Mobile devices. Some devices such as the Motorola A925 support the newer Symbian 7.0 operating system. Support for memory cards is fast growing as a standard as content becomes richer.

5.15 Support for music formats

Other than the well-known MP3, music formats include Sony ATRAC, Apple AAC ,and Microsoft Windows Media. The MP3 format is supported by newer handsets (for example Sony Ericsson P900). Phones like Motorola MPx200 support windows media player. Apple (iPod/iTunes) recently announced a deal with Motorola to license iTunes to Motorola phones. Each format (other than MP3) has its own DRM mechanism

5.16 Support for video

While support for video in 3G devices is expected, many 2.5G devices also support downloadable video (for example Nokia 3660, Sharp GX10). 3G devices such as Motorola A835 support video chat and video streaming. Consider a phone like Samsung SPH-V5400 - a barometer of things to come. Its key feature is a 1.5G hard drive. While 1.5G storage may be possible through other means such as flash cards, the trend appears to making the phone closer to iPod (http://www.apple.com/iPod/) with storage for video clips (as the iPod is for songs). The 'phone', if it may still be called that, also includes a megapixel camera, FM radio and a TV out feature. We see support for video in different formats to be an important trend. TV/Video are seen to be drivers for adoption of 3G. At this point it is difficult to quantify user adoption of video apart from sports and adult content.

Consider the following trends from recent press coverage:

- Broadcasters in South Korea and Japan are launching new satellite broadcasting services that can send video and audio directly to the handheld.
- Prototype terminals for these services are already available
- In the UK, usability trials of multi-channel television to Mobile phones are expected to commence in spring 2005 around Oxford with 500 O2 (www.mmo2.com) customers receiving 16 TV channels including music, sport, news etc.

- The Open Mobile Alliance (OMA) is conducting its own trials with Motorola, NEC, Nokia, Siemens, Sony Ericsson etc.
- MobiTV (www.mobitv.com) has signed up channel partners in the USA for content. It offers a J2ME application for playing video on handsets. Its uptake amongst Mobile operators is still unclear.
- There has been some early success with Mobile TV in Malaysia. It appears the customers want smaller clips rather than live TV. The pricing is also critical. The introductory offer in Malaysia is RM 10 (~$2.60) per month by Maxis and 14.90 (~$3.90) per month by competitor DiGi.(source www.thefeature.com)

Thus, there appears to be a lot of activity around this space but still the customer uptake is unclear.

5.17 Support for Microsoft/Windows formats

We could not find widespread support for Microsoft windows format.

Phones like Motorola MPx200 support windows media formats.

Some Mobile operators are selling branded windows phones. These include phones such as Orange SPV C500, SPV-E200 that support windows smartphone technology. PDAs such as Orange SPV M100 and O2 XDA II support the windows media/pocket PC format.

5.18 Support for MMS

MMS was widely supported through camera phones – either as an inbuilt camera or through an external camera. It's not clear how many of these phones and cameras are actually being used – not many we suspect.

5.19 Support for Bluetooth

Support for Bluetooth surprisingly was not universal. Some high-end devices like Sharp GX20 did not support Bluetooth. It's not very clear why this is the case.

5.20 Miscellaneous

'Swiss army knife' phones such as the rugged Nokia 5410 – complete with a compass, spirit level, thermometer and an optional GPS attachment.

6 Chapter Six: Understanding the Mind of the Mobile Operator

In this section of the book we look more closely at what the Mobile operators are doing within the Mobile development community.

Our central focus is to identify the rationale for the Mobile operators to work with Mobile developers. To be truly successful, we have to understand the mind of the Mobile Operator. We start this chapter by defining some common terminology. This is important since Mobile operators and developers use different sets of terminology, which can cause considerable confusion. Following these definitions we're going to look at the business models and the different approaches to developing business. After considering these approaches, we are going to explore, in some depth, what the Mobile operators are doing within their own developer communities. We are also going to look at some of the ancillary players, who also offer developer programs, in addition, to those provided by the Mobile operators.

The chapter ends by presenting some ideas about how you, an application developer, can gain the attention of the Mobile operators for your own applications. The ideas presented in this chapter are developed in subsequent chapters when we outline the concept of **OpenWaspa** (Open Wireless Application Service Provider Alliance).

6.1 The value chain – detailed view

Before we understand what the Mobile operators want from independent Mobile applications developers, we need to understand how they look at the value chain. In an earlier chapter of the book we presented a high level view, the **Suggested value chain**, figure, below, provides some more detail to that generic explanation. The traditional Mobile operator value chain comprises of the **network equipment, the network operation, terminal equipment or devices and the service providers.** As we move to a more data centric world, three more elements are added to this traditional value chain. These are - **middleware, applications and content**. (Although all these elements arguably existed in their own right before the Mobile data world, those who offer these services have migrated some of their existing business focus to the

Mobile domain to maintain their growth potential, or indeed have created new businesses to exploit this new area). Middleware is made up of two components - these are **service enabling platforms and data middleware platforms.** The service enabling platforms are at the edge of the network. Examples of such 'edge of network service enabling platforms' are: Content repurposing (taking content that was produced for distribution services such as TV or Internet and repurposing it for the size and capability restraints of the Mobile terminal), Personalisation, Real time information streaming, Voice recognition, content management and gaming applications.

Unlike service enabling platforms, which are at the edge of the network, data middleware platforms are inside the network. Examples of such data middleware are: Session management, Optimisation, e-wallet, Security, Messaging, Gateway, Location determination, Protocol conversion and OSS. The service enabling platforms and the data middleware platforms form the modular components - which, when combined with a software layer (application layer), forms a service that the end user finds of value and is willing to pay for.

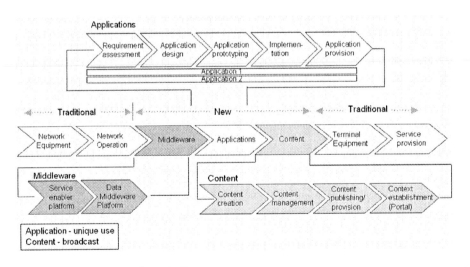

Figure: Suggested value chain

As already stated, the applications layer is software that combines the middleware platforms in a unique way to produce a unique application. From the Mobile operator's perspective, this applications layer is made up of five value-adding components. These components are the **assessment, design, prototyping, implementation and a provision** of an application.

There is an important distinction between application and content. For the purposes of this chapter, **we would define an application as a service that has a unique use.** Conversely **we would define content as something that can be broadcast to a number of users and these users have the same experience.** This is different in case of an application, where each user has a unique experience.

Consider the example of a real time stock information service. Each user will have their own portfolio of stocks and will only want to receive information on their relevant stocks. Users would set up their instance of the Mobile application to provide filtered information to their own stocks; however the service would still broadcast information for all stocks. This makes the application simply a filter for the content, which is broadcast.

Content itself has its own value chain - made up of **creation, management, publishing and often context**. This last part, context, could easily itself be defined as an application. Why is this so? Because context requires knowledge about the user and the location. An application that is able to undertake such tasks will provide the personalised context that makes it all relevant.

We could elaborate the Mobile operator value chain in greater detail but we are more interested in gaining value from it as external developers. Hence, for us, it is more practical to turn the value chain around as shown in the **'Delivery value chain'** figure, below.

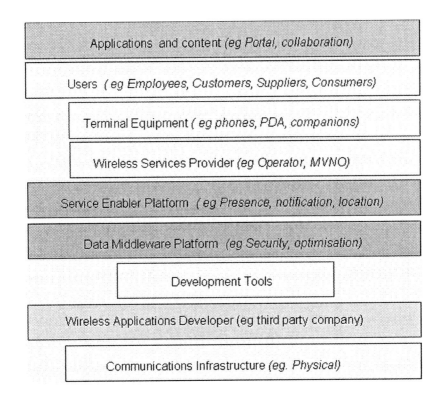

Figure: Delivery value chain

In the delivery value chain, figure, above, which is now in the vertical orientation, we start with the communications infrastructure at the base i.e. the physical stuff that makes up the network. The first block on top of this is the applications developer, closely followed by the development tools. The application developer will use tools to build an application that utilises the service enabling platforms and data middleware platform, which will be unique to the communications infrastructure.

Having built an application, the application itself is offered via a wireless service provider (which in many cases, is the same as the Mobile operator). The user experience comprises of this application running on the terminal equipment.

The reason for presenting it in this vertical format is to show that all aspects must work and interact to produce a successful, useable application. This value chain governs the entire industry, which is depicted in categories as below.

Content Services	Infrastructure	Software and Platforms	Channel and service provision	Wireless enterprise markets
Ringtones, wall paper, traffic, mapping, LBS, Games, Entertainment, media, news, advertising, sports, business news, video and multimedia content	**Network components** network components for EMS,MMS, SMS, GPS GSM, GPRS, 3G **Fixed network components** Components for Provisioning, OSS, BSS, Billing, mediation, WLL,Broadband, Bluetooth **Devices** handsets, PDAs, PANS, special devices, accessories	**Service enablement – edge of** network Content reformatting, Personalization, Real time information streaming, Voice recognition, Content management, Gaming, Data middleware **Service enablement - Inside the network** Session management, m-payment, OTA, Security, Messaging Platforms, Location platforms, OSS **OS** Symbian, Java, Pocket PC, Palm,Brew	Portals, Independent retail outlets, WASP, ISVs System integrators, Consultants, Content aggregators	Enterprise Intranet access, Email access, Sales force automation, CRM, Inventory, Procurement Target markets Travel and transport, Education, Financial Services, Leisure, Manufacturing Government, Retail

6.2 The Mobile Operator's mindset

We have already seen two versions of the value chain above. The Mobile operator sees all applications in the context of this value chain and their mindset is to maximise revenue within this value chain. To understand the mind of the Mobile operator we need to understand their motivations.

In simple terminology: **everything you own they would like to own and exploit for no cost, everything they own, cannot share.**

Whilst this is a little harsh on the Mobile operator it illustrates a point. The point being, that the Mobile operator would like to do everything themselves. However, whilst their ambition is to do everything themselves, they have to find a realistic balance. This balance is presented in '**The balance between Internal and external**' figure.

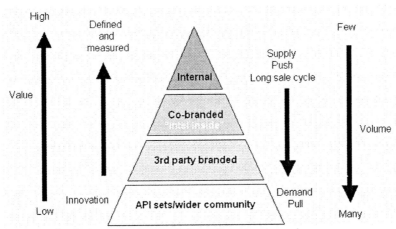

Figure: The balance between internal and external

The Mobile operator's most favoured choice is to develop all applications internally by themselves. However, they have realised that they don't have a monopoly on all innovation. They have also realised that their internally developed applications tend to be late to market and are not as functional as ones developed by a third party application developer.

What the Mobile operators are seeking to do is to find a balance between third party application developers and their own internal developments.

The Mobile operators have realised that innovation produces a very large number of low value applications. Conversely, there are very few well-defined high value applications.

There is also a realisation, within the mind of the Mobile operator that innovation is generated by demand pull i.e. the latest new record produces a new ringtone, game, logo and wallpaper. High value applications tend to require a high degree of market education and as such are supplier pushed, i.e. e-mail on the move, integration into a corporate supply chain management, field service, etc.

Therefore, the Mobile operators need to find a way to balance innovation and value.

They try to find this balance by producing a few of their own internal applications; these are augmented by co-branded and third party branded applications.

However, the highest volume, lowest value, greatest innovations will come from a wide community of developers who produce an application in response to a specific market problem. Creating these applications is the focus of this book.
This type of application will cost very little to build as it will utilise existing services and middleware i.e. not actually build anything new but combine existing components. Hence, they will mirror the 'Leonardo Da Vinci' model.

If we look at the **'Getting more value from the price/demand curve'**, figure, below we get an idea about the motivation of Mobile operators to get access to co-branded and third party branded applications.

This model comes from the software industry and shows how partnerships can be profitable when properly managed.

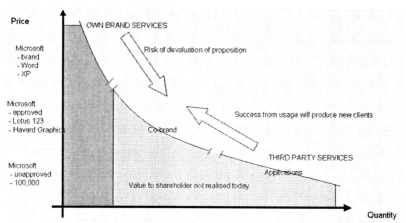

Figure: Getting more value from the price/ demand curve

This diagram shows that there is a lot of value to be gained by being the 'gatekeeper' even though you are not responsible for the marketing or customer service.

In the figure above first shows the economic value that the Mobile operator would achieve by offering only its own branded applications. An analogous example of this from the software industry would be Microsoft's own branded 'Word' or 'Excel'.

What creates further value is being able to co-brand and deliver other applications that your own internal resources have not been able to produce.
Expanding the Microsoft analogy further, co-branded applications would include security or graphics type applications.

In a Mobile operator scenario, third party applications would involve hundreds of thousands of innovative applications that work perfectly well for the user but which may not necessarily be approved by the Mobile operator. The Mobile operators want to gain additional value through deploying a large number of applications but have realised that there are realistic resource constraints, which inhibit their creation 'in house'

Further, Mobile operators have realised, as is the case in the software industry, that you could create value by having a portfolio of applications leading to a competitive advantage.

Coming back to the triangle in '**The balance between Internal and external**', the Mobile Operator is trying to find the balance between their own internal developments (high value, but few in number) and the hundreds of thousands of highly innovative but low value applications.

The trade-off is in the observation that, as they get more companies to develop applications, they risk devaluing their own propositions, but they also stand to gain, since they have invested less of their own resources and are still able to receive revenues from airtime.

Besides this, there is one more element that creates confusion. That element is **the choice of the middleware platform**. The Mobile Operators have to identify which middleware platforms they wish to build themselves and those they would like to acquire from a third party partner. This causes a problem for both the applications developer and the middleware platform developer, since the Mobile operator often attempts to engage with both, using the same developer community.

Indeed, some of the blame lies with the middleware provider as they often write a specific application to show off the middleware platform thus confusing the nature of the sale (i.e. is it about the application or the platform?). However, the application developer and the middleware platform developer are two very different beasts in their requirements, when it comes to the types of contracts and the engagement models that are required to create value.

The 'Developer challenges' figure, below, outlines the challenges for both the application developer and the middleware developer.

The primary concerns for the applications developer is to monetize revenue from the application, show off their innovation, and ensure channel reach and marketing opportunities.

In contrast, the middleware application developer is primarily concerned with technical integration, and exclusivity contract **for their platform and not the 'showcase' application.**

The single largest issue will be the different time scale expected by the parties.

 An application developer would expect an engagement cycle of no more than a few days, whereas the middleware developer would expect an engagement cycle of **nine to fifteen months.**

Hence, application developers who partner with infrastructure developers to sell to the Mobile operators are often disappointed since they have different business models. We explore this dilemma in the next section.

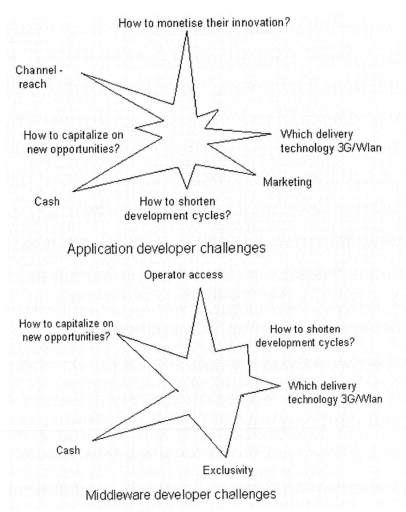

How to monetise their innovation?

Channel - reach

How to capitalize on new opportunities?

Which delivery technology 3G/Wlan

Marketing

Cash

How to shorten development cycles?

Application developer challenges

Operator access

How to capitalize on new opportunities?

How to shorten development cycles?

Which delivery technology 3G/Wlan

Cash

Exclusivity

Middleware developer challenges

Figure: Developer challenges

6.3 Business models

Having understood the Mobile operator's motivations we are now going to look at the business models within the Mobile application landscape. It's not easy to categorise business models. We have defined five types of business models that are prevalent in the market but there are a whole host of grey areas.

The five types of Mobile application business model are: -

- Data enablement
- Wireless enterprise
- Wireless merchant
- Consultancy services
- Content

Data enablement players are players who provide a key technology enabling another application such as the data middleware company. These players principally operate a business-to-business model. Ex www.webraska.com

Wireless enterprise application developers provide a product or service that directly enables an enterprise to use one of its systems or the use of wireless or Mobile technology. www.rim.com

Wireless merchant application developers provide an application to a merchant who provides a service to an end-user. An enterprise application providing the internal back office function. Such a service would allow the merchant to promote their wares over Mobile phones to their customers. www.2ergo.com

Consultancy service companies are very easily defined as a charge or a fee for undertaking work.

Content players develop what the user will actually see. www.iomo.com

The **'Different business models'** figure, below, pictorially shows these business models. We will now have a look at them in a little more detail

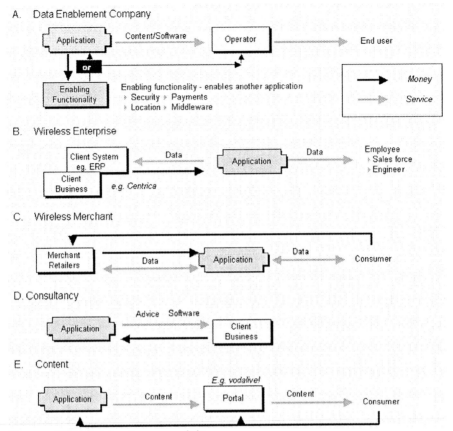

Figure: Different business models

A data enablement company is a company that provides the technology, hardware or software solutions. The principle customer of the data enablement company is the Mobile operator itself. The company will typically sell a solution and they will be paid for this solution by means of a license fee, consulting fees, integration fees, hardware and maybe an ongoing share of revenue that is transacted across their enablement platform. In this case the data enablement company has invested a great deal of time and money in generating and proving that the platform works.

A wireless enterprise is a type of company that provides a wireless solution to an enterprise, which will enable a business process to be used remotely, but connected via wireless technology. In this case the business model is very much like that of the data enablement company, insofar as the types of revenue that can be achieved. The principal difference being that an enterprise developer will not invest time and money into building a platform. Instead they undertake the work in conjunction with their customers, to meet their specific requirements. Unlike pure consulting, the enterprise developer will have role to play in the future delivery and maintenance of the solution.

A wireless merchant delivers a solution to the end user; this end user will be an individual rather than a company. The individual they serve may work for a company, but the individuals themselves subscribe to the service independent of the company's own Mobile data solutions. As with the data enablement companies, the wireless merchant will have invested a significant amount of time and cash in the development of their service, however their compensation for such development will be based on a higher share of the revenue or transactions conducted across their platform. In addition to this they are likely to have a relationship with the end customer. In the case of the data enablement company that end customer relationship is with the Mobile operator.

The consultancy business model, this is the simplest to understand as a consultancy company is paid for the time that they work on a Mobile or project. There is no further ongoing revenue because of their involvement.

In the case of a **content player**, there will be a data enablement company or wireless merchant and/or Mobile operator's internal development platform that is physically between the content and the customer. Whilst the customer believes that they are interacting directly with the content, there are a series of platforms that enable this interaction to actually happen. The content companies themselves are rewarded on a usage basis or on a retainer fee for the continual supply of new and updated content on a daily basis.

Given the variety of business models that companies bring to a Mobile operator and the different motivations, the Mobile operator is not always able to differentiate between the companies that are attempting to knock on its front door.

It is worth looking a little closer at the difference between those companies who wish to incrementally increase the revenues for

Mobile operators and those that increase the cost base of the Mobile operators.

The 'Third party relationships' figure tries to summarise the third party relationships that the Mobile operator has. These relationships can be classified into three broad groups.

The first is **strategic relationships**; this could include corporate venturing, alliances, partnerships, research and development.

The second category is **supply relationships**; these are outsourcing, system integration and consulting technology supply relationships.

The third category is **third party developers. This is the category that actually increases the revenue of the Mobile operator, whereas the other two have a burden on cost.**

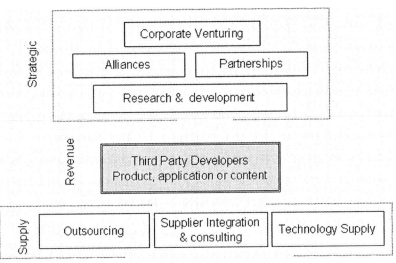

Figure: Third party relationships

There are potentially three avenues to interface with the Mobile operator – the **commercial team** (headed by the commercial director), the **technology team** (headed by the CTO) and the **Mobile operator's own developer program** (if one exists). These are usually held by different board/ executive members within the Mobile operator's organisation and these three are often in conflict.

Both developers and larger technology suppliers approach the Mobile operator as external suppliers.

The technology suppliers to the Mobile operator wish to differentiate themselves when being considered for new projects. One of the mechanisms they have for creating differentiation is to bring third party companies to their bid (just like the middleware platform providers before).

What we mean by this is that the technology supplier such as Nortel, Nokia, Motorola will be actively going out to source interesting applications that show their platforms or their technologies in better light. The supplier will then bundle the value of these applications or services within the sale of their own platform technology.

Typically the CTO will own the relationships with these technology suppliers, and the CTO sees their remit as bringing additional applications to the table.

However, this will be in conflict with the commercial director who will own the strategic relationships. The commercial director will see their role as business development and that they should own prime responsibility for bringing additional applications to the Mobile operator that increase the value of their organisation. They see that they are able to fulfil this duty by asking their strategic partners to find and source interesting third party application companies, which will also increase their value and relationship with the Mobile operator.

Thus, it is very likely that the CTO and commercial director will source similar types of applications from different companies on different commercial terms through different partners, which will confuse the decision-making process.

Then, there is the third avenue; i.e. the Mobile operator's developer community that is also busy trying to source applications. The third party developer program, which is run by the Mobile operator itself, sees its value as sourcing third party applications.

The board director who owns the responsibility for this unit will be arguing that the technology supply, which is coming via the CTO, should not be bundling additional applications into their bid, as this only confuses what the technology supplier will actually deliver for the price that they are offering.

In the same way, this director will also be arguing with the commercial director against using their partnerships to source third party applications. The rationale for argument being the premise that their own internal development program will be best suited to identify the applications that the Mobile operator needs rather than a third party company attempting to find interesting applications that the customers of the Mobile operator are not demanding.

What we can conclude from our third party relationship discussion is that it is very unclear who actually owns third party relationships with the Mobile operator. It is further clear that they themselves don't really know who owns the overall relationships. We could call this predicament as the Mobile operator's dilemma!

Thus, while a significant majority of the Mobile operators and the technology suppliers are building their own developer forums, in the hope that they will be able to supply interesting applications, they are fundamentally in conflict.

However, what we, as developers, now need to determine is, given the business model we have, which are the best partners to go and try to work with?

6.4 The developer's dilemma

Rather like the old game the prisoner's dilemma, we have a four box quadrant that provides a difficult decision making scenario. The figure below shows us the dilemma. As a Mobile applications development company, whom do you turn to, to work with and partner with, to create wealth in the fastest possible time?

There is a need to work with the network equipment manufacturers, as ultimately the service will be delivered on their technological equipment. There is also a need to work with the middleware companies, as without these, services will not be enabled. Further one must work with the terminal equipment manufacturers, because unless your application is designed with terminal equipment in mind, the user experience is likely to be very poor. The final component is the service provider, as without a relationship with the service provider (Mobile operator) there is no mechanism for revenue collection.

This presupposes that as a developer you are going to utilise existing network, middleware and terminal equipment resources. Should your business model be the middleware developer then there is a different dilemma, as you are actually the enabler, but you would still need close relationships with the other interested parties.

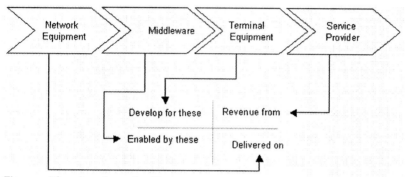

Figure: The developer's dilemma

There is no easy decision, hence the reason for having a dilemma.

But the key issue for a developer is where do you find information?

The developer needs information to ensure that their application for technology will provide the user experience, for which, the user is willing to pay.

The '**where to go for information**' figure highlights the source of information that the developer needs. Some of this information depends on the type of application that the developer is building.

From a network equipment manufacturer, a developer may need access to network upgrades and improvement data. Without this you would be unable to guarantee that your application will continue to work as the network is improved and upgraded. Access to information from the middleware company is required as there is a great deal of specific functionality and integration required to ensure that there is scalability to the application. It is self evident that information is required for the terminal equipment, but with over 450 unique devices now available (each device varies by OS and functionality) the amount of information is too vast for many companies. Without information on the terminal equipment, you are unable to ensure that the user experience will be pleasurable and whether the user would return to reuse your application, hence creating an ongoing revenue stream.

Each Mobile operator has, unfortunately, implemented and built their network differently from other service providers. Thus, when you've finally built your application for one Mobile operator, it is unlikely to work on somebody else's different network implementation.

The implication of this is that additional resources are needed to make sure that your application works on different Mobile operator's networks. If you are unable to get your application working on all Mobile operators' networks, you will be unable to access the viral effect of one user telling another user about your application.

Without this viral effect, you'll be dependent on huge marketing budgets to be able to build users and revenue for your application.

The primary problem within this area is where to find all the necessary information that you need to build a workable, sustainable and maintainable application.

It is an easy question; however there doesn't appear to be a single resource available today.

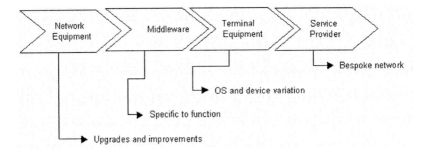

Figure: Where to go for information?

Yet, if we look to a different market for example the PC market, there is an abundance of information. Why is this market so very different?

Fundamentally, the PC market has a vertically integrated and controlled development process, whilst this is a simplification of the reality from the developers perspective it is reasonably true. This is indicated in the **'Control' figure**, below.

From the chipset, through to the OS, through the applications layer, the developer environment and the operating environment, there are a very few, very powerful players. These players work very closely together to ensure that their systems and process work together and that all the necessary information is available.

Whilst some readers may argue that not all information is available on the PC platform (such as source code not being known in all cases), nevertheless the vertical value chain is complete (even if imperfect in some cases). This analogy to the PC market is not designed to be exact. It is only to illustrate how organised this market is, even with these difficulties, if compared to the Mobile market.

We believe the best way to describe the Mobile developer environment is fiefdoms or cartels. We would not go so far as to suggest collusion amongst the players, but rather that this should serve as an example that describes a point. The point being, why it is so difficult to get hold of the relevant information and build Mobile applications that work on all networks and all terminal devices?

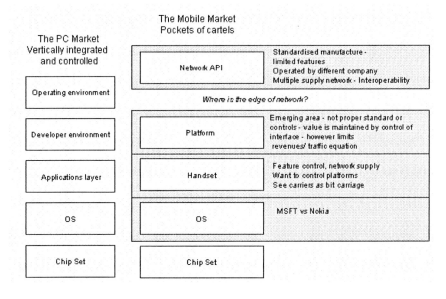

Figure: control

When looking at the figure above, one can see that in the Mobile domain, unlike the PC domain, there is no integrated and controlled developer process. In a Mobile domain each critical component has very large and dominant players who would like to control much more than they actually can.

One of the significant issues is - where is the edge of the network? This issue in itself could form a book. The key point here being that there is no single source of information for the Mobile application developer. Given that there is no single source of information, but there is wide number of Mobile developer communities which one should you work with?

6.5 Developer programs reviewed

The review of the developer programs is broken into two stages.

The first stage tries to explain the different motivations that lie behind developer communities, when viewed from a Mobile operator's perspective, the second stage then puts the different developer community is in two categories based on these motivations. Some of the key developer communities are reviewed to provide some information to help you decide whom you would like to engage with.

Like the buyers warning - caveat emptor (the buyer beware) the reader should understand that these developer communities continually change and the information presented has a finite life.

Hence, we have provided a link on the open gardens web site to update information about developer communities www.opengarden.com/developerupdates.html

The first stage we're going to look at in this comparison of developer community is the motivations and objectives that lie behind them.

The first consideration is to decide the potential points at which a Mobile operator can engage with a third party since as developers, we must fit into one of them.

The **'where to engage'** figure, overviews the potential points at which a Mobile operator can engage with a third party.

The **first point** of engagement is where there is no more than an idea. The difficulty for Mobile operators engaging with developers at the idea stage is – Mobile operators have no skillsets/knowledgebase to pick out the stars from the duds. Indeed, if you only have an idea, taking it to a Mobile operator is not likely to produce any results at all. .

The **second point** of engagement is when the application developer has produced a proof of concept, the idea has some validation and there is some numbers, which demonstrate the magnitude of the opportunity. It is unlikely that there would be a different response at this stage from the idea stage as there is still no prototype that would demonstrate the reality of the application.

The **third point** of engagement is when the prototype has been built and is ready for testing. At this point the Mobile application developer has used development tools to produce something that can be shown and therefore interacted with. The difficulty now for the Mobile operator is choice. What we mean by choice at this point is - which handset, which platform and which network would it work best on? From the experiences of existing, applications developers, even at this stage, the results are not good because Mobile operators are still unable to determine which applications will make a difference.

The **fourth point** of engagement is when the prototype has been built into a product this product has been integrated and is capable of being tested on a live network. The key issue here is that the developer has now committed a high degree of real cash.

Given this high degree of commitment from the application developer, the application developer hopes that the Mobile operator should be able to provide some cash to cover the cost of proving and integrating their application into the Mobile operator's network.

However, the Mobile operator sees it differently. They believe that the potential opportunity for the Mobile applications developer is so large that they (the developer!) should be willing to fund the cost.

The Mobile operator doesn't always appreciate that the applications developer will not have significant reserve funds to enable this to happen – given that no revenue is coming in. Given this chasm between the two, often there is no progress made with a loss to both parties.

The **fifth point** of engagement is where the application developer, to overcome the problem of engagement at point four, has funded the integration and testing. At this point the Mobile developer wishes to have a contract with an underlying commitment written in. The application developer justifies such commitment on the premise that they have funded to date all of the costs of integration development and testing. All that remains is the marketing of the application to generate revenue. The application developer sees the Mobile operator as that marketing channel.

Yet again, the Mobile operator doesn't see themselves as a marketing channel at all! Furthermore, it isn't known which application is going to be successful, so one will not be marketed in favour of another.

The Mobile operators have provided commitments previously to smaller third party companies, which have unfortunately backfired and have cost them a significant amount of money due to under performance of the application. They are very cautious about commitments.

As with engagement point 4, point 5 has the same problem, in so much as, that there's a chasm.

The **sixth point** of engagement is where the Mobile application developer has crossed the chasm of point 4 and 5, and has a successful application that is in the market. However the application developer now finds that there are several other developers who have produced a functionally similar application! The Mobile operator wishes to source the best application.

The Mobile operator now enters into an evaluation and testing phase of several competitive but similar applications. The application developer at this point must provide a differential advantage to be selected. This differential advantage could be the cost, the user experience or the lack of integration hurdles for example. If selected at this point the Mobile application developer should enjoy a very high margin. However, the Mobile operator still determines the overall margin that can be achieved. The margin that the Mobile operator often gives does not justify the overall investment to get an applications developer to this point.

The **seventh point** of engagement, the last point, is where an application developer has an application in the market that is generating revenue. They take this application to a competitor Mobile operator in the aim of gaining additional revenue by having a second or third channel to market. It is interesting because at this point the application developer is now able to force the margin to issue in their flavour as they have a successful application generating income with known user volumes to negotiate with.

At this point in the development and engagement cycle the Mobile operators are at their weakest as they have a requirement, but the risks for the Mobile application developer are the highest, because if they don't get beyond this point they are unlikely to survive.

What is interesting is that the Mobile operators perceive that there are so many Mobile application developers that they are not worried if **one or two thousand die** on route.

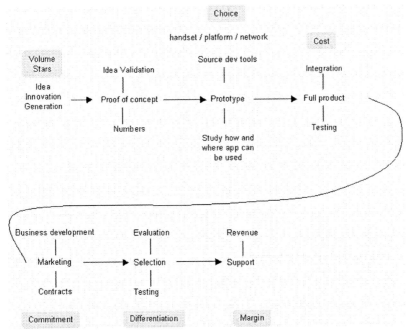

Figure: where to engage

What we can understand from the seven points of engagement is that the objectives vary.

If the objective of the Mobile operator is to provide differentiation, they can afford to engage later in the cycle. If their objective is to find innovation they have to engage earlier in the cycle. What is evident from the discussions in the opening parts of this chapter is that Mobile operators see that internal developments should provide differentiation and third party companies should provide innovation, as this would allow them to balance the resources needed. However, when we consider where third party application companies come into the Mobile operator, at the strategic, technical or applications level, we find that they'd do not have these clear demarcations between differentiation and innovation.

(The above paragraph is worth re reading again since it offers some crucial insights).

Increasingly, the criteria that the Mobile operators are setting imply that they only want to engage with application developers who are at the end of the cycle.

Currently, we observe base level criteria such as:

- Must be standalone
- Will only require the network to provide carriage
- Must already have a commercial customer
- Must be of industrial grade
- Technology assessment is simple
- Has a proven business ROI

The result of these criteria means that the hurdle rate is so high that they would produce no inputs, and input in this case means a company that has developed a Mobile application.

However, if the Mobile operator lowers these hurdle rates then they may find that they are in inundated with very novel ideas but no real applications that they can take to market. If they have too many novel ideas they are unable to determine if any of them have any value. The volume is too high and the output to zero. There is therefore a careful balancing act between getting the criteria right.

Therefore a Mobile operator, who has a business unit that is able to make quick decisions on its input, could be a Mobile operator who may succeed in bringing great applications to market. However the quickest decision is not to proceed with anything!

Therefore, the Mobile operator needs to be able to make an assessment of their own engagement processes. We believe that there are four scenarios for engagement process. These scenarios will be underpinned by **some winning characteristics**, these characteristics would be; **speed of decision-making, focus, a structured approach and sufficient resources to actually make something happen**. The converse of these could be described, as **losing characteristics** that would be; slowness to commit, unclear position in the market, lack of focus and inconsistent behaviour.

Sadly, today many of the Mobile operators tend to exhibit the losing characteristics, whereas some of the technology providers/ system providers, who run their own developer forums, tend to display winning characteristics!

This becomes increasingly obvious when we look in detail at the differences between the developer communities provided by the Mobile operator, the technology providers and the system solution providers.

If you look at the Mobile operators themselves, there are four scenarios.

- Scenario a, is where the Mobile operator is a follower
- Scenario b, is where the Mobile operator is focused
- Scenario c, is where the Mobile operator wants to be the leader in Mobile data and finally
- Scenario d. is where they do nothing.

Looking at each of these in turn:-

Scenario a, is where the Mobile operator wants to be a follower, by this we mean that they would exhibit the following characteristics,

- Work only with developed applications (working, live applications)
- Turn away non working applications
- Provision of carriage only for the application – no need to do any integration
- No differentiation in service created by applications compared to others (i.e. an application of similar standing is already in the market)
- Low risk to build applications revenues
- Integrated into a wider partner program
- Very clear external communications to manage expectations and resources

An example of a follower would be Nextel in the U.S. whilst they are a leader in revenue from Mobile data, they will only engage with applications that work. Even though they engage late in the process of development - they were the first to provide applications and they now hold a leadership position in terms of revenue and the number of applications they are able to serve.

Within this area of being a follower there should be two further categorisations - that of the active follower and the passive follower. The active follower, who is working with an application provider, will ask their partners to bring to them new applications. In this case the Mobile operator may choose to work with a partner on applications that don't yet meet the characteristics as described above.

Passive followers conversely follow tightly and rigidly the characteristics above. Even an existing application provider, who has an existing profitable application with the Mobile operator, would have to wait until an application is ready for launch rather than finding that the Mobile operator is willing to start to work together to get the application to market earlier.

Scenario b, is where the Mobile operator wants to be focused, by this we mean that they would exhibit the following characteristics:

- Willing to work with applications in the development
- Be very selective in focus
- Provide solutions
- Unlikely to have any differential offerings
- Provide high technical resources to help and assist partners in integration and delivery
- Integrated into a wider partner program
- At this stage there a few Mobile operators who are displaying this characteristic, however technical players such as Blackberry would be a good example of a Mobile focused developer community.

Scenario c, is where the Mobile operator wants to be a leader, by this we mean that they would exhibit the following characteristics:

- Be willing to build an aggressive developer community
- Provide a high degree of developer support
- Will publish a high degree of API and SDK programs pre launch
- Will never second guess which ones will be successful
- Will be seen as an innovator with an innovative brand
- An example of a Mobile operator who wanted to be a leader was 02, in that they developed a program, SourceO2, however it is apparent that more recently they have been moving towards being a follower of scenario a.

Scenario d, is where the Mobile operator decides to do nothing in the area of engaging third party application developers. Their belief is that there is not sufficient value to be created in this area at this time, and that their time would be better spent developing their own applications internally rather than engaging third parties. The best example of this will be T-Mobile.

Summary of engagement scenarios

The motivations and commitments required by the leader are very different from those of follower. The primary difference will be at the engagement stage. The leader will engage earlier in the cycle than the follower. The biggest difference this means for the Mobile operator is the resource required, both at the engagement point and with respect to the resources and facilities needed to enable developers to work with the technology.

6.6 Under the microscope

The Mobile operators compared

Some developer programs run by Mobile operators are compared as follows:

SourceO2: www.sourceo2.com Provides a one-stop shop for developers to work with www.mmo2.com. Source O2 is a pioneering program with events, knowledge share, test beds, platforms etc
It has 15,000 plus registered users and started off as wanting to be the leader in wireless applications. The program included developer labs, which were closed down for lack of interest. Through their revolution program, they offer distribution opportunities for applications.

Vodafone: http://www.via.vodafone.com/ 'Via Vodafone' recently revamped Vodafone developer program provides access to the Vodafone network.

Orange: http://developers.orange.com Orange developer program - Another recently re invigorated program provides access to the Orange network

The system partners compared

A comparison between some system partner development programs is as below:
Blackberry: www.blackberry.com/developers

Due to the large number of Blackberry devices, this program is increasingly attractive to developers. Blackberry development is done in Java, which makes the application relatively easy to develop.

Power by hand: http://www.powerbyhand.com/
The program aims to distribute applications through their Mobile operator customers. Members gain access to technical information and a platform, which enables them to get simple, downloadable applications to market without the Mobile operator.

Java One: http://java.sun.com/javaone/index.jsp
Enables global development of Java software similar to the Microsoft developer program. Provisions tools and knowledge to enable developers to implement and test Java applications. However, note that merely conformance to Sun's Java standards is not enough i.e. a Mobile operator may have their own tests to conform to before the application/game is accepted.

Nokia: www.forum.nokia.com
Forum Nokia provides a single stop shop to create applications for Nokia devices. It also allows access to the customer through Nokia's own distribution program. Seen within the industry as a competitor to the Mobile operators due to its own distribution program i.e. brand.

HP (Hewlett Packard) Mobile regional services bazaar: www.hpbazaar.com
HP service implemented as through a series of regional programs and partners as a place to nurture innovation. Application developers can gain from HP's distribution, sales and marketing programs.

6.7 Concluding remarks

In this chapter we have studied some of the motivations and objectives of the Mobile operators' developer communities and tried to marry these to the requirements of application developers.

In doing so we hope that we have highlighted some of the issues and concerns from both parties and from both points of view.

What we have realised is that process is broken, insofar as the Mobile operator's resources and developer's requirements are still some distance apart.

It is evident that there is a party needed who can bridge this gap.

We do not believe that the WASP (Wireless Application Service Provider is the best solution) to bridge this gap. We believe this because the WASP does not benefit both sides of the equation but rather introduces an additional element that is attempting to produce economic return for its shareholders.

In doing so it is reducing the margin and hence the survivability of its partners above it and below it in the food chain.

We believe that an open Wireless Application Service Provider Association, OpenWaspa is needed as previously described. We describe this concept in the next section

Having said this it is good that both the Mobile operators and others are providing facilities and resources to be able to bring Mobile data applications to market. They have different strengths and different focus and we hope that you now have better information and are able to engage with the right partner.

The market faces much change and consolidation in the coming months. We will be tracking ongoing changes at this site www.opengardens.com/developers.

In this section, we have seen the flaws in the existing ecosystem. In the next chapter, we shall see how these drawbacks are likely to be overcome.

7 Chapter Seven: Open Gardens Revisited

7.1 Background

All of our discussions so far have been in preparation for this chapter. In the last chapter, we saw 'How Mobile operators currently think'. This thinking is currently changing. In this chapter, we see how this evolution affects applications development. Here, we bring together all the concepts we have discussed with the goal of seeking solutions to the problems facing us.

Knowing what we know now, we can look at OpenGardens from a different perspective. In a nutshell, OpenGardens is discussing the problem of 'access to distributor channels'. The Mobile operator is the largest and potentially the most lucrative distributor channel. So far, we have seen viewpoints from both the Mobile operator and the developer/innovator. In this last chapter, we are going to look at the specific strategies i.e. what could we do now - from a developer's perspective.

Clearly, a 'big bang' opening of walled gardens is not likely to happen. Hence, we look at two strategies – the initial phased opening of walled gardens followed by a world of 'ideal OpenGardens'. Ultimately, a 'nirvana' environment of OpenGardens will enable the creation of many complex, interconnected applications such as the 'air grafitti' application. That world may not be so distant in the future considering the rapid advance of so many technologies since the rise of the Internet within the last decade. However, it will arrive incrementally.

Often, the industry takes a one-dimensional view of 'open' vs. 'closed' and its impact on Mobile operators and developers. But, it's more complex than that for a number of reasons –

a) There are external factors that come into play, for example regulatory/legislative issues.
b) The channel (operators) themselves are changing. A 'phased opening' of gardens is happening even now which creates opportunities for us even when we do not have open APIs.

c)   A web services/open API initiative is already under way potentially leading to a full OpenGardens vision. We discuss the web services model under the SDP concept below.

d)   The problem of incremental revenue shares in the value chain that affects developers. We introduce a new model called OpenWaspa below that addresses this issue.

This section may be summarised in four broad subject headings:
   a)   Discussion of 'external factors' that influence Open Gardens
   b)   A phased opening of walled gardens
   c)   The 'ultimate' vision of OpenGardens including web services and the opportunities that are created by it
   d)   The OpenWaspa model

Before we start discussing the OpenGardens concepts, we need to briefly discuss what OpenGardens is not. OpenGardens is not the same as open source. Similarly, OpenGardens is not related **only** to 'DRM' i.e. Digital rights management/copyright protection. Recently, some analysts have raised the spectre of a malevolent 'Napster-like' (www.napster.com) scenario where new smartphones pose a threat to content. The protection, or otherwise, of content is a different issue from the one we are discussing here (although DRM has an impact on OpenGardens and innovators as we see below). For starters, in general, we are discussing services and not content. PC connectivity, memory cards, Bluetooth, Infrared can all be used to copy content – but this issue is distinct from the one we are discussing. For the record, we do not believe that opening up the device is a threat to content protection in itself.

7.2 External factors that influence walled gardens

There are some factors which are neither controlled by the Mobile operator, nor the developer, but have a bearing on Mobile services. We must not forget that the telecoms environment has only just been recently deregulated. There are still many areas which are unclear. When viewed globally, these issues are not easy to manage. We give some examples below and discuss the impact of these factors on Mobile services and walled gardens.

7.2.1 Legal and regulatory issues

Mobile operators are now faced with extra statutory responsibilities when it comes to the Mobile Internet. These include – monitoring for SPAM, managing DRM (Digital Rights Management), monitoring adult content etc. Some examples of the legal and regulatory issues include:

a) Monitoring SPAM when it arises from outside a country. The same applies to more malicious schemes – such as fraud using premium rate text messages many of which originate from outside a given country.

b) Definition of 'legal' content that varies according to geography – For example in Scandinavia – the rules for adult content are relatively less restrictive whereas in other countries – there are severe restrictions on content. In some countries, there are extra religious restrictions – for example 'astrology and predictions' are restricted due to religious reasons.

c) The Mobile operator is now tasked with managing DRM for content providers. Content providers are understandably suspicious of another 'Napster' on the Mobile Internet. But DRM can get very complex and comes with a cost of enforcement. For example – true protection of content would entail controlling all means of data transfer – including Bluetooth, extendable memory devices etc. Specific technologies have additional considerations. For example – because MMS follows the MIME (email) standard, it is possible to separately save components of an MMS message just as we can in an email. Preventing this requires more management/technology and hence more cost.

d) Content classification – especially ensuring that adult content is not accessible by children.

e) Mobile SPAM – Right from the outset of Mobile services, advertisers have been drooling about the mythical 'Starbucks' application. In a nutshell the application is supposed to work like this - As soon as you pass a café, say Starbucks – you were supposed to be sent a 'coupon' for 10% off your next cup of coffee. The implementation of this service is not trivial and the application itself is not cost effective – at least for now. However, it illustrates the point that –

given a free rein, advertisers would bombard us with messages. And this does not include the illegal operators who would send much more SPAM through schemes with potentially illegal connotations.

f) Because the Mobile Internet pertains to IP connected devices, it is likely that the viruses, which reside on the Internet, could take on a Mobile flavour.

g) Privacy and consumer protection including issues like use of camera phones, management of use profiles (i.e. who can see your profile), privacy in relation to location based services which we discussed before etc.

h) Data protection and consumer rights – For example how will the data collected on movement of customers be managed? Who will have access to it and under what circumstances? What information will be stored on local devices as opposed to the server? How can the data be protected at the device level?

i) Inclusive design to enable access for all customers including the elderly and the visually impaired. Mobile operators like Cingular Wireless in the USA have introduced new applications that respond to spoken commands using voice recognition technology to cater for the visually impaired. Inclusive design is becoming mandatory. In many countries, new legislation is coming into force to cater for people with disabilities. For example, in October 2004, UK introduced the Disability Discrimination[16] Act

These issues have two common elements – firstly they are new and because there is no precedent there is an element of ambiguity in their enforcement and secondly – the operator is often at the forefront of any enforcement.

We see a number of initiatives being implemented industry wide. For example UK Mobile operators announced a body called the Independent Mobile Classification Body (IMCB) that will regulate and classify adult content on Mobile phones. It is an independent subsidiary of the premium rate regulator, ICSTIS (http://www.icstis.org.uk/). Operators will create filters compatible to recommendations of this body so as to restrict access to adult content especially to protect children from

---

[16] http://www.legislation.hmso.gov.uk/acts/acts1995/95050--c.htm.

unsuitable material. The classification framework will be compatible to other media like cinema and games. Prior to this body, one operator (Vodafone) had implemented its own classification scheme. This cased considerable resentment in the industry with a Mobile operator playing the role of a moral enforcer. Hence, a need for an independent body was recognised.

Similarly, the Mobile Marketing Association has been formed as an industry body to monitor Mobile marketing http://www.mmaglobal.co.uk/. However, the critical fact remains that the Mobile operator is the major enforcer of these policies even when they are created through industry consensus or government regulation – hence the impact on walled gardens because these policies cannot be enforced without a measure of control. A 'free for all' would result in more harm than good to the industry's reputation as a whole.

7.2.2 Industry maturity or smoke screen?

Is the above discussion a sign of industry maturity or is it a smoke screen to keep profit margins high? There is no doubt that some element of control is beneficial to the industry and can only lead to a more prosperous future. It is designed to overcome the worst excesses of the Internet for example SPAM, preventing children from accessing adult content, copyright violation etc. Mobile operators have a large part to play in enforcing these controls. In many cases, they simply have no choice – the government or regulatory bodies create legislation that makes it mandatory for the Mobile operator in turn to create restrictions.

However, the ancillary question is – how 'inclusive' are these proposals to the garage developer/innovator? The more inclusive they are, the better it is for the whole industry. Once guidelines come into place, how will these guidelines be enforced on sites not managed by the operator? Will small service providers be able to join regulatory bodies or would the barriers to membership be placed so high (for example, high membership fees) that small providers cannot join? The answer is not clear but the issue is a significant influencer in the Open Gardens debate. Creating barriers for pure commercial reasons is simply not sustainable. For example, the first 3G operator in the UK ('3' www.three.co.uk) does not permit open access to the

Internet. But, the second 3G operator to launch (Vodafone) www.vodafone.co.uk **does** allow open access to the Internet.

'3' itself is now considering tiered portals like I-mode. Hence, it is 'opening up'. To 3's credit, it is a pioneer and being a pioneer means facing the unknown – hence a walled garden may actually have benefited '3' at least in understanding the market when no prior statistics are available. As expected, like AOL, restrictions will melt away over time as the business matures for everyone.

Thus, even within the Mobile operator community, competitive pressures will bring down the barriers. And to top it all, completely novel ideas can turn up which change the industry model inside out.

Consider the case of Xingtone (www.xingtone.com) - which is what we call a 'disruptive application'. The company offers an application that lets users make their own ringtones (from music they already own). If downloads are an indication, customers want this software. Curiously, one of the recent investors in Xingtone is Siemens (www.siemens.com) - a handset manufacturer!

7.3   Phased opening of the walls

The metaphor of a walled garden creates a mind-frame of restrictions and 'only one way out'. It causes us to lose sight of other avenues, which could be lucrative even today. Historically, the Mobile operator portal has been the main focus of a walled garden. The relative listing on the portal along with a billing system and knowledge of the user's location are the three main elements of the garden. From an application perspective, ultimately we need open APIs for a true 'level playing field'. However, open APIs are a component of open gardens and not an end in itself. Even if full openness is not available, we could adopt some strategies in the interim taking advantage of the phased opening of gardens, which is already happening as below. We recommend a combination of the following strategies – all of which can be deployed today.

7.3.1 Off-portal revenues

The whole objective of a portal is to maximize exposure to the target customer base. The end game is to get access to a large number of customers, across multiple operators globally and manage the relationship securely from a single point. Portals are not the only means to achieve the goal of maximum exposure. Off portal revenue models provide a mechanism that enables smaller players to access a large customer base.

One example live and thriving today is a service called Bango (www.bango.net) whose platform achieves this goal for the Mobile Internet. The key element is Bango's numerous relationships with operators around the world. Bango operates in 130 countries and gives access to a user base of millions of users. As per their web site at any time, Bango handles over 450,000 ringtones, 120,000 Java games, 80,000 videos and 250,000 images through its platform.

This makes a truly open, global platform that allows content providers access to a large customer base. Since the platform handles billing and security, it provides a 'one stop shop' for a new service that can work across operators. At its basic level, Bango is like 'Worldpay' (www.worldpay.com) on the Internet but by providing cross operator access, Bango has the potential to truly create a business model for small providers – starting from the smallest 'rowing club' to a large media company

7.3.2 Wholesale models MVNOs

We have discussed MVNOs previously. MVNOs broaden the market and provide more channels to market for the developer.

7.3.3 Developer programs

Developer programs can act as a channel to market since many developer programs have a 'marketplace' that enables the developer to sell their application. Further, some programs offer a tiered model where you can list your application on a second tier portal for a much lower fee. If it is successful at that stage, you can take it to the top tier portal for a higher exposure. This two-stage model gives more players a chance to enter the market.

7.3.4 Cross-industry examples

Cross industry initiatives like SMS short codes also help to increase market exposure i.e. the same service can be accessed by customers from more than one operator. In the UK, the best example of this initiative is the 'five digit short codes'. These are common across the networks and being only five digits - are easier for customers to use. They have been introduced only in

Dec 2002 and are now already a common feature across the industry. Their ease and capacity to manage high throughput means that they have led to a class of new services like interactive TV programs such as "Big Brother".

7.3.5 Degree of branding control

If you have a brand that is a leader in a specific sector, you have a much higher negotiating capacity since you have the potential to drive traffic to the portal as well as gain a share of the revenue. Thus, a brand like 'Russell Grant' for astrology www.russellgrant.com has considerable leverage in portal positioning although it is not in the same scale as a 'Coke/Disney' brand. We have also seen the impact of MVNO/wholesale models where many brands are launching new services leveraging their existing strengths. In this case, the branding is reversed with the MVNO brand being at the forefront and the Mobile operator brand at the background. An example of this relationship is Virgin Mobile (www.virginMobile.com) and T-Mobile (http://www.t-Mobile.com/) in the UK and USA.

7.3.6 Portal positioning and access

We have discussed previously that portal positioning is key to exposing your service to new customers. Many operators are now introducing tiered portals. The OpenWaspa model listed below also leans towards a tiered portal. Thus, there are opportunities even if your service does not get top listing in the main portal but manages a good listing on a secondary portal from where it could graduate to a top tier portal. Recently, we have seen the growth of Java portals (as opposed to WAP portals) and they are significant because many brands are likely to launch Java portals – thereby opening up more channels to market. Some operators like Orange (www.orange.com) in Europe have actually opened up their WAP portals and allow developers to use their Orange 'Kiosk' billing system assuming that they meet certain criteria.

7.4 OpenGardens and open APIs

In the previous sections, we saw the phased opening of walls and the impact of external factors. Ultimately, developers seek open APIs. From this section onwards, we will discuss the impact and benefits of open Mobile telecoms APIs.

We can illustrate the impact of open APIs from an application Ajit was involved in creating. The basic idea was to create an application named 'Rich Maps'. The concept was seemingly simple and also helpful for the end user.

Take a feed from a mapping company, overlay information from content companies (cinemas, superstores, train stations etc.) and serve the resultant 'Rich Map' to the user's handheld device. The application functions similar to 'find my nearest' but on a browsing model. For example, if I'm standing at Oxford Circus tube station in London, I can 'navigate' the streets around the station on my handheld device 'browsing' both inside and outside the shops on these streets.

In other words, I would have 'visibility' both outside and inside all shops on streets such as Regent Street

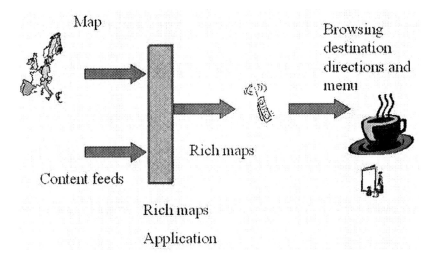

However, this application has many practical hurdles to cross:

- How do we sign up the 'point of interest' on the map?
- How can we make it easy for the vendor to sign up?
- How many such providers will the vendor sign up with? Each Mobile operator, other startups, etc.
- How accurately can we determine location?
- Can the same application be deployed across all Mobile operators?
- As high street shops change over time, how will we keep the information up to date?

- Taking this a step further, how can we browse inside a shop? For example, viewing a restaurant menu.
- How will users discover the application?
- What would make this a valuable application to end-users and to customers?
- Would this application be fundable?
- Is this one application or a federation of applications acting together?
- How can we manage payments?
- How 'Mobile' is this application? Considering the 5 'M's, how much information can be acquired by other means? For example, browsing the web from home.
- How much of this information can be acquired by browsing the web from the handheld device i.e. still getting the information from the device 'on the move' but not from our application?

In fact, the problems are not new. We have touched on these problems before in discussing the failed Paybox model.

Notes:
- For the purposes of our discussion, we have ignored technical problems like rendering maps on different devices
- The application combines more than one element and is a complex, utility based application
- The client side access method is important (BREW, J2ME, Browsing, etc.) but access to the data/services is critical. There will be many clients with varying technologies. The data/service will 'make or break' the application.
- The vendor's dilemma (where a vendor is defined as any point of interest on a map ranging from the corner grocery shop to the museum) is the first hurdle to cross. How many providers will the vendor sign up with?
- This problem is larger than 'content aggregation'. In fact, it is 'service aggregation'. There are many players in the content aggregation market such as 'yellow pages' and startups like www.e-street.com. A content aggregation model requires that the content be up to date, which quite often, it isn't. Further, there are costs associated with collection and aggregation of content, all of which add to the eventual cost of the service. Finally, content aggregation often doesn't cover details of a specific establishment (e.g., restaurant menus)

- The application would benefit from a feature of web services called 'late discovery and binding'. In traditional applications, the interface between two services is agreed in advance. Web services, on the other hand, are loosely coupled. The actual implementation is known only at the last minute (thereby ensuring that the information is up to date).

7.5 OpenGardens and the SDP concept

In spite of the problems currently facing the development of applications such as 'Rich Maps', times are changing because the Telecoms application landscape itself is changing. Mobile operators are likely to change only because it's in their own interest to do so and there is a need, which they cannot fulfill. In the above 'Rich maps' example, the application cannot be developed 'top down'. It has to be developed in an organic manner that is closer to a set of web pages/web services rather than a single monolithic application. **No company, no matter how well funded, can achieve this on its own.**

Only by 'opening up' the telecoms network, can this class of application be developed, i.e., API enabling the telecoms platform.

In our view, the model/infrastructure we discuss below is the one most likely to emerge in most Western/Asian geographical locations. Indeed, it's not revolutionary but evolutionary. Companies such as IBM are already advertising 'On Demand' computing. A recent issue of business week (www.businessweek.com) discussed these principles in the article 'Tearing Down the Walls in Telecom' Business week March 4 2004.
http://www.businessweek.com/technology/content/mar2004/tc2004032_4116_tc076.htm

Whilst we have talked of the need to open up the telecoms network and the current state of Mobile data applications development, we have not yet seen **how** the telecoms network would be API enabled. For this we have to switch hats and cross over to the telecoms network – specifically in the understanding of the SDP (service delivery platform) concept.

The **SDP (Service Delivery Platform)** is the software element within the Mobile operator's infrastructure that manages the creation and deployment of the next generation voice and data services. The actual features of the SDP depend on the vendors who implement it. Most large Telecoms vendors such as Ericsson, Logica, IBM etc. have SDP solutions. Many Mobile operators are already considering/deploying these solutions. There have been previous attempts to achieve the same goals as the SDP, notably through the Intelligent Network concepts, but these were limited. The Internet, XML and web services standards existing today make the SDP approach much more feasible than those attempted before.

7.6 The SDP Infrastructure

7.6.1 Navigating uncharted territory

The SDP presents technical mechanisms to 'API enable' a telecoms network. However, the Telecoms industry faces a number of non-technical issues in its attempts to 'open up'. This is unfamiliar territory for the industry. Prior to SMS, the industry had many price plans, but only one product, voice calls.

Compare this with a large supermarket stocking around 150,000 items across 150 categories. These items are all predictable (and have been so for around 75 years). In contrast, services that the Mobile operator is now encountering are new and unproved. Even if they are branded, their future is uncertain and unknown.

In the existing telecoms ecosystem, due to high infrastructure costs, regulatory pressures security etc. Mobile operators tend to deploy only a small number of proven applications over large consumer markets to ensure that they recover their costs. In contrast, in the emerging ecosystem, the Mobile operators, are faced with a large number of applications spanning boundaries across partners. Mobile operators are swamped with large number of unproved developers and unproved applications.

It is also unclear as to how these new applications will be supported. After the cost of the network itself, customer service is the next highest cost for many Mobile operators. With the new applications, the need for customer service and quality of service becomes even greater. Traditionally, in enterprise or Web based consumer applications, the Systems Integrators (SIs), Value Added Re-sellers (VARs) and Independent Software Vendors

(ISVs) have provided consultancy, support and customer service. However, SIs, VARs and ISVs are not used to Telecoms specific issues such as roaming. Hence, support for new applications needs careful thought from both the Mobile operators and their potential partners, such as ISVs.

In this context, the conservative approach taken by the Mobile operators should come as no surprise in these economically uncertain times.

To protect themselves from unknown issues in uncertain economic climates, Mobile operators have continued to adopt the existing model i.e. 'build it in-house/no partners' approach better known as the familiar 'walled garden'. This model works well with existing services. However, as growth from traditional avenues decline and new services emerge, there is growing realization that the model needs to change.

7.6.2 Abstracting the telecoms network

As we have seen before, the SDP is the software element within the Mobile operator's infrastructure that manages the creation and deployment of next generation voice and data services.

The SDP provides an ecosystem for the creation and deployment of services. It sits at the cusp between the core network and IT services. Within the Mobile operator's infrastructure, it falls within the IT services rather than the core network but integrates with network elements such as WAP servers, SMSCs, positioning servers etc. Thus, the SDP concept brings together two distinct worlds, the world of computing with an emphasis on aspects such as databases, computation etc. and the world of telecoms, with an emphasis on issues such as latency and reliability of connections and so on.

The SDP's function is service provisioning. Its success lies in abstracting the telecom's network to services created by third party developers. **Crucially, this abstraction has to be both commercial and technical.** Technical abstraction, to hide the complexities of the telecom's network to developers and at the same time protect the network from security threats AND commercial abstraction, because small developers cannot negotiate complex commercial agreements across many geographical locations/Mobile operators.

From a technical standpoint, the environment must appear to be similar to existing technologies used by the third party developers. It must also facilitate similar development cycles such as build, test, deploy (or build, deploy, test☺) followed by commercialization. Developer tools such as the SDK (Software Development Kit), IDE (Integrated Development Environment), along with emulators, test platforms etc. are all necessary to make it easier for developers to create and deploy applications. Besides these, telecom's specific aspects, such as Identity management etc., should be available via simple APIs to third party developers. We expect tools vendors like MS Visual Studio, Borland JBuilder, SunONE Studio BEA Workshop and others to provide tools to create SDP applications.

However, the bigger need is commercial feasibility. The developer must have a single point of contact so that their application can be profitably deployed over a wide customer base. This 'one stop window' must also manage commercial relationships and the revenue/billing. Today, in the industry, only few players like Qualcomm/BREW have both the technical infrastructure and commercial agreements in place. Following Qualcomm's example, we expect more major players like IBM to enter this space providing both the technical tool sets and the commercial relationships.

7.6.3 Technical abstraction

A block diagram of the SDP is as shown below

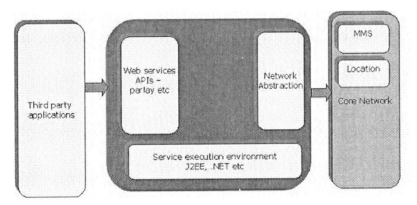

The main functions of the SDP platform are to:

a)   **Abstract the network layer** by providing interfaces to network elements
b)   Provide a **service development and execution environment**
c)   Provide a **set of web services APIs** that are accessible by third party developers

And optionally, a **content delivery platform** for provisioning multimedia content

In addition, the SDP also provides ancillary business and technical management utility services like partner management, billing, load balancing, redundancy management etc. to services created by third party developers.

As developers, we are primarily concerned with the 'Service evelopment and execution' function where we create a new service using technologies like J2EE or .NET. Services need simple and secure access to the lower network elements. The SDP provides this function through a set of low-level software calls, which abstract the core network functionality. Calls to the core network elements are implemented using gateways and standardised technologies like web services. Finally, once the service is created, it needs to be 'exposed' to third parties. This function constitutes the 'API enabling' the telecoms network which we have been discussing right from the outset. The role or service exposure is performed by web service based standards like Parlay X, which are explained below.

Let's pause for a minute to gauge the significance of the above section.

The SDP platform allows a third party developer to create a new service running inside the Mobile operator's network. This service has access to low level APIs via the network abstraction layer. In addition, the service can be exposed to external parties. If there are no restrictions (or only reasonable restrictions such as security), this constitutes the breaking of the walled garden.

The actual standards and the level of exposure depend on the individual Mobile operator. However, Mobile operators who have installed/considering the installation of an SDP platform at least have the technical capability to open up their network.

The adoption of SDP and other standards like Parlay is happening currently. According to www.parlay.org as at May 2004, four Mobile operators have plans for commercial deployment in 2004. These are BT, Colombia Moviles, Orange/France Telecom and Sprint. Because the SDP platform lies at the intersection of telecoms and IT, there are a range of players who are developing SDP platforms. These include IT vendors (BEA/Sun), Telecoms equipment manufacturers (like Ericsson) and new specialist software vendors like Aepona (www.aepona.com)

7.6.4 Standards

What standards/technologies could apply in the task of opening up a telecoms network?

One body addressing this problem is the Parlay group (www.parlay.org). Parlay is a standards body and it was the first body dedicated to the task of opening up the telecoms network. It has made some progress but has also had some setbacks.

It's important to note that the task of API enabling a telecoms network can be performed independent of a standards body like Parlay. In fact, parlay has already had a few setbacks as we see below – but the basic approach is valid. Once a Mobile operator decides to open up, they could do so using their own API. We expect that they will be 'compatible with Parlay' but with proprietary extensions.

To explain this, consider the analogy of SQL standards in the relational database environment. SQL (Structured Query Language) is standardised via ANSI – American National Standards Institute (http://www.ansi.org/). SQL from all database vendors such as Oracle (www.oracle.com) is ANSI compatible. However, vendors have added their own proprietary extensions. That's just how the industry works, and we expect the same with all standards. However, the standards themselves are important as a lowest common denominator, irrespective of how they are implemented by specific vendors.

There are a number of standards/initiatives in this space – with some overlap between them.
- **Parlay:** As we mentioned above, parlay is the original standard formed solely for the purpose of opening up the telecoms network.

- **Parlay-X:** Parlay X was proposed in May 2004 to overcome the drawbacks of Parlay. Parlay X is based on web services.
- **OMA:** The OMA is proposing it's own standards in this space and there is discussion to align these standards with Parlay
- **JAIN:** JAIN is a Java based standard/environment, which extends the Java environment to the SDP. Think of JAIN as extending the Java space 'above' J2EE (i.e. J2ME, J2SE, J2EE and then JAIN).

These standards and technologies apply to different facets of the SDP.

J2EE and Microsoft .NET are dominant in the service development and execution phase. Another option at the service development and execution phase is JAIN, which is built solely for the purpose of service creation within the telecoms environment.

The Parlay standards address the issue of service exposure. Parlay was first introduced in the year 2000. Two things happened since then that affected their uptake – firstly the global economic downturn that hit the marketplace and secondly the initial reliance of parlay standards on the CORBA (Common Object Request Broker Architecture) architecture. CORBA was relatively complex to implement, which hindered the initial technical uptake of Parlay. With the advent of J2EE and web services, Parlay X was released in May 2004. Parlay X overcomes the limitations of Parlay and is closely aligned to web services. This makes Parlay X a much better standard for developing open telecoms applications.

According to the Parlay 4.0 Parlay X Web Services Specification (v 1.0.1 June 2004)

The Parlay APIs are designed to enable creation of telephony applications as well as to "telecom-enable" IT applications. IT developers, who develop and deploy applications outside the traditional telecommunications network space and business model, are viewed as crucial for creating a dramatic whole-market growth in next generation applications, services and networks.

The Parlay X Web Services are intended to stimulate the development of next generation network applications by developers in the IT community who are not necessarily experts in telephony or telecommunications. The selection of Web Services should be driven by commercial utility and not necessarily by technical elegance. The goal is to define a set of powerful yet simple, highly abstracted, imaginative, telecommunications capabilities that developers in the IT community can both quickly comprehend and use to generate new, innovative applications.

Each Parlay X Web Service should be abstracted from the set of telecommunications capabilities exposed by the Parlay APIs, but may also expose related capabilities that are not currently supported in the Parlay APIs where there are compelling reasons.

Thus, the Parlay X APIs are built on top of the Parlay APIs. The relationship between Parlay X and Parlay is as shown below

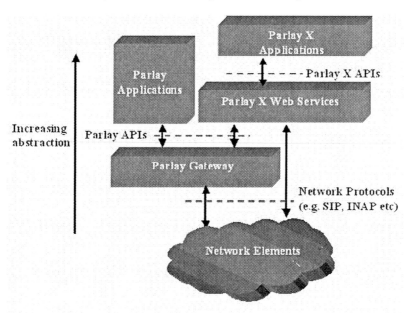

Source: Parlay 4.0 Parlay X Web Services Specification (v 1.0.1 June 2004)

7.6.5 Parlay X APIs

As developers, Parlay X APIs give us the opportunity to create interesting applications. Hence, it's important to know which

APIs are being released first and where to look for planned APIs in the future.

The APIs outlined below are part of the Parlay 4.0 Parlay X Web Services Specification (v 1.0.1 June 2004)

7.6.5.1 Third party calls

This API initiates a call between two parties depending on the occurrence of a threshold situation for example between a client and their broker when a stock quote reaches a certain threshold.

7.6.5.2 Network initiated third party calls

Handles scenarios such as automatic handling of calls if the subscriber is not available (such as routing the call to a secretary)

7.6.5.3 SMS

The parlay/OSA web services standardise calls to the SMSC for sending and receiving SMS messages. Currently these calls are proprietary (i.e. specific to the SMSC vendor).

7.6.5.4 MMS

Similar to SMS, Parlay/OSA web services standardise calls to the MMSC. Functionally, this is similar to the MM7 interface we discussed before.

7.6.5.5 Payment

The payment APIs used by the Parlay group are the same as those created by Paycircle (www.paycircle.org), the relationship is not clear to us. APIs support payment reservation, pre-paid payments and post-paid payments.

7.6.5.6 Account management

For pre pay customers, the Account Management APIs support account querying, direct recharging and recharging through vouchers.

7.6.5.7 User status

The User Status APIs are used to get the status of the user and are used in applications like Buddy lists.

7.6.5.8 Terminal location

As its name suggests, the terminal location API is used to determine the user's location.

7.7 The OpenWaspa model

In the previous sections, we discussed the SDP from a technical perspective. Here, we look at the commercial implications of the SDP business models. Strictly - the Mobile data industry does not include corporate applications (i.e. applications deployed within the corporate firewall). Corporate applications have different revenue model (i.e. a software sales model). Hence, we face a typical scenario of a consumer facing application where applications need high volumes, have low unit margins and long sales cycles to sign-up agreements with partners. In a consumer market, applications must be scaleable and have the capacity to address a large customer base.

The existing commercial model (which we describe as the 'waterfall model') is doomed to fail in the long run. The waterfall model starts with the developer and ends with the customer via the Mobile operator. **With each player in the middle asking for their own margin!**

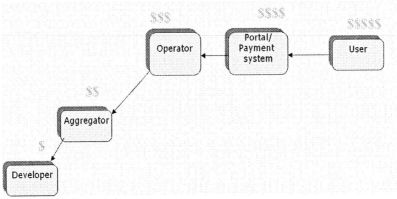

The waterfall model

The 'waterfall model' has some flaws, which the industry often overlooks. The bottom of the food chain (the plankton – if you take a marine biology example) is expected to feed the 'whales' (i.e. the Mobile operators at the top of the food chain). But, the food chain can only work if the 'plankton' i.e. developers survive. By the time the meagre revenue trickles down to the developers, they do not have a viable business. There is little revenue at the top of the food chain either i.e. this is still an emerging industry with considerable fragmentation. So, margin aside, there is little revenue in the first place – leading to a vicious circle.

The middlemen serve the Mobile operators at the expense of the developers. Again, the developer is affected. In our view, the waterfall model (and other variants such as the aggregator model where every middleman takes a portion of the revenue) are adopted from more mature industries and simply bolted down to the Mobile data industry (which is an emerging industry).

It is wishful thinking but it does not work. The problem is depicted below

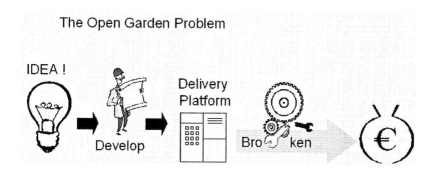

The alternative model - which we describe here as the OpenGardens model - is not novel, in fact this model is already hugely successful at Amazon, EBay and so on.

In a nutshell:

- Make it cheap and easy for the small developers to plug directly into the network infrastructure
- The big players (Amazon, EBay, Mobile operators) provide the hub/bazaar/marketplace for a small portion of the revenue and a small fixed fee
- Eliminate the middleman using the Internet

The OpenGardens model has the following characteristics:

a) **Open telecoms networks:** i.e. third parties can create and run services on the Mobile operator platform with no restrictions (other than security etc.).

b) **Low fixed price connectivity:** It must be easy for developers to 'plug in'. We propose a **fixed fee say $100/year** to plug in

c) **No middlemen!!**

d) **Upgrade possibilities:** Possibilities for developers to upgrade i.e. if a Mobile operator likes the application, they do direct deals with the developer

e) **Complex applications:** Created through a combination of multiple APIs.

This model would create a virtuous circle rather than a vicious one. We believe that either a progressive Mobile operator or new entrants (such as MVNOs – Mobile Virtual Network Operators) would adopt this model.

It would lead to OpenWaspa – Open Wireless Application Service Provider Alliance rather than the WASP (wireless applications service providers). **The crucial difference between the WASP and OpenWaspa is the absence of a revenue share at each stage. The OpenWaspa simply has a small, fixed fee allowing the players to 'play' and the Mobile operators to choose the best players.**

The OpenWaspa model is shown below:

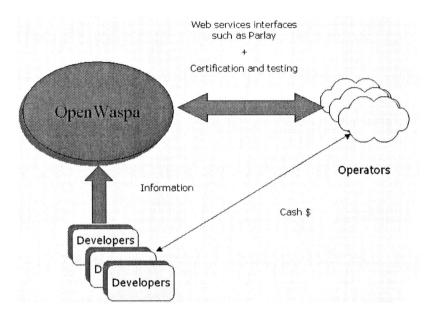

So, what kind of world will this be? The answer is not far from us. Like the Internet, there will be some big winners and many sites (web pages) – all useful to users. Venture capitalists will fund some big players but the greatest impact will be at the grassroots level.

This would also overcome many of the problems Mobile operators face (for example the dilemma between differentiation and innovation).

We conclude this chapter with the following observations:

a)   The phased opening of the gardens is creating opportunities for innovators

b)   The ultimate vision of OpenGardens i.e. Open APIs will create a new class of applications where innovators/developers are integral to the whole ecosystem and

c)   Mobile operators are addressing the technological problems by deploying SDP platforms. The bigger commercial problem for the 'garage' developers can be addressed by using the OpenGardens model as outlined in the five points above. This model could help unleash the creativity and innovation which hampers the industry today

We leave this chapter with the question – 'Is there any evidence **today** of deployment of applications similar to the 'Rich Maps' application? Or, is this simply too futuristic? We can share two examples – both in the Mobile multiplayer games space and both in Japan. The first is a Mobile multiplayer game called the 'Mogi' game (www.mogimogi.com). The Mogi game is a hybrid treasure hunt game that combines both the Internet and a Mobile location based service. Players can collect and trade based on real life locations. Another example is Samurai Romanasque game from Dwango http://www.dwango.co.jp/ . The gameplay of Samurai Romanesque includes factors such as the weather, which are based on an actual weather feed. While these examples are rudimentary, they are nevertheless pioneering and a reflection of true mobility.

Conclusions

We have come a long way – starting from our original question – **Why don't a thousand flowers bloom on the Mobile Internet?** And it's two ancillary questions:

- How can we foster innovation within the Mobile data industry?
  And,
- How can we combine multiple services, bodies of knowledge and trends to create a commercially successful Mobile data application?

We have seen the innovator's dilemma, the developer's dilemma and seen that even the Mobile operators have their own dilemma!

But this is an industry approaching a trillion dollars (that's a 1000 billion dollars!!) globally with 1.5 billion customers worldwide – a quarter of the world's population.

That's cause for optimism! We see opportunities in this industry for innovation. The phased opening of walled gardens followed by possibilities of more complex applications in future have the potential to truly transform the commercial landscape.

OpenGardens acts as a catalyst – we don't know how you can use the foundations we have laid out here, but we are keen to know. We look forward to hearing from you through our web site and blog on www.opengardens.net

As the old Chinese proverb goes – 'May you live in interesting times'. We certainly are!

References

1) Leading the revolution www.leadingtherevolution.com - Gary Hamel
2) Next generation wireless applications – www.paulgolding.info - Paul Golding
3) www.wikipedia.org
4) www.java.sun.com
5) www.thefeature.com
6) Loosely coupled – by Doug Kaye www.rds.com
7) Messaging Applications – Ajit Jaokar and Russell Buckley www.messaging-applications.com
8) The Innovator's dilemma – Clayton Christensen http://www.claytonchristensen.com/ '
9) The Innovator's Solution - Clayton Christensen http://www.claytonchristensen.com/
10) Mobile Digital rights management – Ajit Jaokar www.futuretext.com
11) Competitive Analysis – Michael Porter http://dor.hbs.edu/fi_redirect.jhtml?facInfo=pub&facEmId=mporter&loc=extn
12) Competitive Advantage – Michael Porter - http://dor.hbs.edu/fi_redirect.jhtml?facInfo=pub&facEmId=mporter&loc=extn
13) Moriana group - http://www.morianagroup.com/
14) Services for UMTS – Tomi Ahonen and Joe Barett http://www.tomiahonen.com/
15) I-mode strategy – Takeshi Natsuno
16) www.mgain.org
17) www.ovum.com European MVNO market: the Ovum perspective *By: Keshinee Shah, Senior Consultant, Wireless Consulting Practice*
18) http://www.adl.com/ 'Mobile Virtual Network Operators' Fraser Curley Bonn November 2001
19) MVNO explosion – paradigm shift in the making ATKearney.

About the authors

Ajit Jaokar

With a unique perspective spanning multiple aspects of Mobile Applications, Ajit is an innovator and a pioneer in the Mobile Data industry.

Ajit has been involved in developing and deploying Mobile applications within communities since its early days. He was first exposed to SMS (text messaging) applications through a UK based company, which deployed SMS based alerts to customers in major UK shopping malls. Ajit was involved in understanding the community, the requirements of the retailers and the customers in formulating a strategy to introduce new services within the community (payments, multimedia, location etc.).

Ajit has published a number of reports on topics such as MMS (Multimedia Messaging), DRM (Digital Rights Management) and worked with leading edge clients in Europe.

His current area of focus includes Mobile gaming, Multiplayer Mobile gaming and Bluetooth communities. His reports have been used by companies such as Nokia, '3' etc.

Ajit speaks at Mobile related conferences - both on strategy and technology. Currently, Ajit plays an advisory role to a number of Mobile start-ups in the UK and runs Mobile communities within the UK especially the Mobile Applications club at http://www.ecademy.com/module.php?mod=club&c=24 .

Over the last year, the club has spanned more than 26 countries and features members from all over the world (from USA, UK, Europe and even to places like Siberia and the Seychelles).The club has attracted the innovator / early adopter – who believe in Mobile technology, business and innovation despite the doom and gloom of the dot com industry around them. Ajit's open and flexible approach has led to a two-way flow of knowledge and insights.

Ajit empathises with developers and OpenGardens is a personal mission for him.

Ajit has an Electronics engineering degree from the University of Bombay.

Tony Fish

Tony has been involved for 20 years in the Mobile, wireless, telecom and satellite industries.

Tony divides his time between his non-exec roles at 2Ergo (an AIM listed Mobile application company); Chronos Technology Ltd. (supplier of Sync solutions to Mobile operators); Abrocour (a leading supplier of Smart Home technology); Subzone (the first Mobile DIVA – Direct and Interactive Value Acquirer) and his role as founder and CEO of AMF Ventures, a leading strategic advisory and investment company, exclusively focused exclusively on the wireless industry with its headquarters in London.

Tony is known for his innovative approach, strategic and economic insight, analysing, matching and executing merger and acquisition activities within blue chip corporations as well as entrepreneurship and shrewd business decisions with regard to early stage businesses and their growth. Tony has been associated with high tech investing since his first IPO in 1994.

Tony currently chairs the world's largest user wireless group www.wireless.ecademy.com This group has over 3,500 members and is sponsored by Intel, BT, Motorola, Swisscom, T-Mobile, The Cloud, Fujitsu Siemens, NetGear, Palm and Sun.

Tony regularly speaks at conferences on the financial and economic impact of wireless and Mobile in our lives and how to generate wealth from participating in probably the most exciting industry.

Tony has a B-Eng in Electronics and Electrical Engineering from the University of Reading, and an MBA from Bradford School of Management. Is a member of IEE and CIM.

www.tonyfish.com

Printed in the United States
29076LVS00002B/15